CW01583445

Enacting Nature

Ecocritical Perspectives on Indigenous Performance

P.I.E. Peter Lang
Bruxelles · Bern · Berlin · Frankfurt am Main · New York · Oxford · Wien

Birgit Däwes and Marc Maufort (eds.)

Enacting Nature

Ecocritical Perspectives on Indigenous Performance

"Dramaturgies"
No. 33

The book was subject to a double blind refereeing process.

No part of this book may be reproduced in any form, by print, photocopy, microfilm or any other means, without prior written permission from the publisher. All rights reserved.

© P.I.E. PETER LANG S.A.
Éditions scientifiques internationales
Brussels, 2014
1 avenue Maurice, B-1050 Brussels, Belgium
info@peterlang.com; www.peterlang.com

ISSN 1376-3199
ISBN 978-2-87574-146-2
D/2014/5678/35

Printed in Germany

CIP available from the British Library and from the Library of Congress, USA.

Bibliographic information published by "Die Deutsche Nationalbibliothek".

"Die Deutsche Nationalbibliothek" lists this publication in the "Deutsche National-bibliografie"; detailed bibliographic data is available on the Internet at <http://dnb.de>.

Table of Contents

Acknowledgments

The editors would wish to acknowledge the assistance of Audrey Louck and Gregory Watson, the series' editorial assistants, who helped with bibliographical and house style formatting. The editorial team is much indebted to Victoria Kramb for her meticulous proofreading. The P.I.E. Peter Lang production team, in particular Alice de Patoul, provided useful advice all along the editorial process. Most importantly, the editors wish to extend their thanks to Ms. Émilie Menz, P.I.E. Peter Lang's Senior Editor, for her dedication to this project.

Editorial Policy

The editors have decided not to italicize Aboriginal English and Māori words so as to acknowledge their integration in contemporary Australian and Aotearoa/New Zealand language.

A Chorus of Ecological Voices

Enacting Nature in Contemporary Indigenous Performance

Birgit Däwes and Marc Maufort

Universität Wien and *Université Libre de Bruxelles (ULB)*

Ecological strategies, sustainable solutions for environmental change, and their effective communication across various cultural boundaries are among the primary challenges of the twenty-first century, joining human beings of diverse origins in a planetary community. For the development of policies to counter climate change, and to protect and share resources globally, it is crucial to establish and exchange, across regional differences and national boundaries, interculturally sensitive understandings of specific places—including the spatial, temporal, physical, social, and cultural coordinates that shape them. In the humanities, a reading practice has emerged over the past few decades that highlights the close connections between nature and culture, and which addresses some of these questions. Ecocriticism, famously defined by Cheryll Glotfelty as "the study of the relationship between literature and the physical environment" (xviii), has developed into a popular and widespread methodology (cf. Heise, Garrard, Goodbody and Rigby). Its manifestations include a wide spectrum of critical stances and practices, from activism to systems theory, and from thematic analyses of nature writing (Payne) and "the environmental imagination" (Buell *The Environmental Imagination* and *The Future of Environmental Criticism*) to the dehumanized, ecocentric revisions of "deep ecology" (Naess, Sessions) and material feminism (Alaimo).

At the intersections between Indigenous Studies and ecocriticism, however, critics have only recently begun to explore the ethnically and/or culturally specific implications of nature, environment, and the land. In the field of postcolonial studies, Cara Cilano and Elizabeth DeLoughrey have inquired into "global knowledges" from an ecocritical point of view, and Graham Huggan and Helen Tiffin combine, in their volume *Postcolonial Ecocriticism,* both these methodologies in order "to contest [and] to provide viable alternatives to western ideologies of development" (27). The project to develop new parameters of thinking about ecological imperialism, environmental racism, and the ecological implications of cultural diversity and mobility was also continued by Elizabeth DeLoughrey and George B. Handley in *Postcolonial Ecologies: Literatures of the Environment.* While a small number of Indigenous texts play a role in these studies, the coordinates for a fruitful critical investment in representations of nature are still being mapped. Common stereotypes, such as the wilderness topos, the "ecological Indian," or the keeper of a planetary spirituality, have proven tenacious and difficult to overcome. In North America, for instance, the predominant image of Indigenous people has been linked to Romantic discourses of nature: Indigenous people are presented as noble custodians of their environment; instinctively harmonious with their environment, yet unable to meet the challenges of technology and civilization. "Time and again," Shepard Krech writes in a seminal study in 1999, "the dominant image is of the Indian in nature who understands the systemic consequences of his actions, feels deep sympathy with all living forms, and takes steps to conserve [...] the earth's harmonies" (21). Along with these simplifying stereotypes, the American continent was inaccurately described as an Edenic "wilderness" by European settlers. This situation was quite similar in Australia and New Zealand: the depiction of the "new" territory's population as nature's "children" implied static and primitive cultures, fitting perfectly into the patronizing paradigms that justified colonization.[1] The images of the eco-Indian or eco-Aboriginal—effectively revived toward the end of the twentieth century—have been powerful instruments of dispossession and displacement.

Ecocriticism thus requires not only historically specific and culturally sensitive investigations of the "connections between social injustices and environmental degradation" (Adamson 20), but also a critical awareness of the underlying discourses, power structures, and ideologies in representing these injustices. Ethnic studies, postcolonial studies, critical race theory, and Indigenous studies provide helpful frameworks for what Joni Adamson projects as "a more inclusive environmentalism and a more multicultural ecocriticism" (xix). Furthermore, Franca Bellarsi, in "Ecocriticism, or Beyond Eurocentric and Anthropocentric Anti-Humanism," calls not only for redefinitions of what classifies as "nature writing" but also for the development of "an ecocritical aesthetics that would acknowledge biodiversity and the dynamically interrelated processes of nature, as well as recognize the limitations of human

[1] For more detail about the conflicts between settlers and Aborigines in Australia, the reader could turn to Russel Ward's *Concise History of Australia* (in particular to the chapter entitled "New Settlements and New Pastures," 86-138). For an account of how settlers viewed Māori, the reader can consult Keith Sinclair's *A History of New Zealand* ("Australian Colony" 29-50, and "Brave New World," 51-70).

language when it comes to expressing the non-human realm" (164). In spite of the fast expanding field of ecocritical practice, therefore, much remains to be done at the theoretical sites of discussion.

In terms of genre, ecocritical practice has been predominantly concerned with fictional and non-fictional prose texts. Relatively few approaches have been taken toward the analysis of drama and theatre from an ecocritical perspective. Among those, one can list Gabriel Egan's monograph on *Green Shakespeare*, Baz Kershaw's *Theatre Ecology. Envronments and Performance Events*, Wendy Arons and Theresa J. May's edited *Readings in Performance and Ecology*, a supplement of the *Journal of Dramatic Theory and Criticism* (Richmond), a special section of *Theatre Topics* on "Performance and Ecology" (Arons), a special issue of *Canadian Theatre Review* entitled *Theatre in an Age of Eco-Crisis* (Nelson Gray and Sheila Rabillard), as well as a few scattered articles. However, no systematic approach has yet been undertaken to explore the interconnections between ecocritical methodologies and Indigenous theatre, drama and performance.[2] This volume hopes to serve as a first comparative exploration of these combined fields; bringing together analyses of Indigenous plays from North America, Australia, New Zealand and the Pacific Islands. Together, the contributions that have been collected here seek to explore the relationship between Indigenous drama and the "environment" in the widest sense—as place, land, nature, wilderness, social space, "thirdspace" (in both Soja's or Bhabha's senses[3]), and "alterNative" space. Our notion of ecocriticism is not limited to environmentalism as a form of creative advocacy, but it acknowledges, in its basic assumption, Robert M. Nelson's insight that "cultural identities, like individual identities, emerge not from class struggle but rather from the land" (7). This anthology does not purport to offer a definitive theoretical model through which ecological Indigenous performance must be examined. Rather, our approach implies an exploration of the multi-faceted languages of ecology on the contemporary Indigenous stage.

The contributors to this anthology make abundantly clear that the link between Native people and their environment, while certainly evading Euro-centric values of materialistic exploitation, does not imply a kind

[2] The October 2013 issue of *Theatre Journal*, edited by Ric Knowles, is devoted to "Interspecies Performance." While this thematic perspective is not directly related to ecocritcism in performance, it certainly prolongs debates initiated by earlier work on ecology in theatre and drama (Knowles).

[3] Next to Bhabha's famous post-colonial denotation of the "third space" that results from hybridity, in which negotiations of new stances become possible, Soja defines "Thirdspace as an-Other way of understanding and acting to change the spatiality of human life, a distinct mode of critical spatial awareness that is appropriate to the new scope and significance being brought about in the rebalanced trialectices of spatiality-historicality-sociality" (57).

of easily homogenized and unproblematic harmony. Viewed together, the essays published here debunk the myth of the "Ecological Indian," which Greg Garrard describes as follows: "Since the sixteenth century at least, 'primitive' people have been represented as dwelling in harmony with nature, sustaining one of the most widespread and seductive myths of the non-European 'Other'" (Garrard 129). By contrast, the essays assembled in this collection examine in detail how the Indigenous "Other" relates to specific environments in various parts of the globe, focusing on different Aboriginal nations and tribes. At the outset, one limitation must be acknowledged: this comparative perspective focuses on Indigenous works written mostly in English or (in one case) in French, i.e. languages of the colonizer. Nonetheless, in many of the works analyzed, linguistic and cultural palimpsests of oppressed Indigenous cultures can be detected, either in unglossed scenes in the Native tongue or in the use of Aboriginal aesthetic devices. Thus, the corpus of works dealt with here is often hybrid, an in-between-ness that enables the authors to forge unique ecological discourses. The approach adopted in this volume seeks to open up perspectives which will have to be extended to the works of artists working in non-Western languages.

The various contributions to this book, whether focusing on North America or Oceania, offer insights into diverse performance practices, including drama, dance, and film. Most of the essays alternate between issues of ecojustice and of what could be termed ecospirituality. For our purposes, ecojustice is understood as the intersection between ecocritical discourse and "political, ethical and religious overtones" (Clark 120). By analogy, we construe ecospirituality as that aspect of ecocriticism dealing with the spiritual bond between Indigenous people and the Earth.[4] In their exploration of these notions, the essays collected here survey a large corpus of work by such established Indigenous artists as Tomson Highway, Drew Hayden Taylor, Marie Clements, Yvette Nolan, Kevin Loring, Wesley Enoch, Hone Kouka, Briar Grace-Smith and Witi Ihimaera, while shedding light on emerging figures as well.

The collection's opening essay, Birgit Däwes's "Stages of Resilience: Heteroholistic Environments in Plays by Marie Clements and Yvette Nolan," mirrors the agenda of most of the essays gathered in the volume. Unwilling to "hug trees," Däwes develops a model she calls heteroholistic cultural ecology. Her concept is derived from Hubert Zapf's notion of "cultural ecology," which allows the critic to see literature "as a specifically powerful form of symbolically staging and

4 The double focus of this anthology on the notions of ecojustice and ecospirituality owes much to productive discussions about ecocitrical theories with Franca Bellarsi, to whom the editors extend their thanks.

exploring the interrelationship between cultural and natural processes and energies" (Zapf 49). It also echoes Cherokee scholar Billy J. Stratton's "heteroholism," which Däwes praises for its acknowledgment of "the importance of land and local landscape to Indigenous cultures without denying the dynamic, in-progress, and unstable characteristics of these cultures" (28). Such a perspective enables Däwes to examine how the stereotypical notion of the "Eco-Indian" is debunked in experimental plays by Marie Clements (*The Edward Curtis Project*) and Yvette Nolan (*The Unplugging*). Däwes's contribution is followed by a series of articles focusing on North American material. Ric Knowles's "Mounds, Earthworks, Side Show Freaks and Circus Injuns" goes back to the roots of Indigenous performance in North America. Knowles argues that ancient Indigenous mounds and earthworks can be regarded as literal performances expressing a special relationship with Native ecospheres. Knowles further describes his involvement as a dramaturge in an ongoing theatre project devised with LeAnne Howe and Monique Mojica, entitled *Side Show Freaks and Circus Injuns*. The latter attempts to provide a contemporary equivalent to ancient mounds performances. In the next essay, Jaye T. Darby focuses on *Out of Dust*, a little-known play by an artist of Cherokee heritage, Lynn Riggs, first produced in 1949. While this work focuses on European settlers, it nonetheless evokes issues connected with Native ecology: because they view the land as a purely materialistic entity devoid of any spiritual dimension, these European characters ultimately wreak only havoc. Moving to the contemporary period, Maryann Henck examines how Drew Hayden Taylor's *The Berlin Blues* enlists the use of Native humor in order to oppose the tendency, only too evident in today's globalized world, to ghettoize Native culture through ecotourism. The play satirizes a European attempt to commodify Ojibway culture through a theme park display. In contrast to this form of cultural imperialism, Taylor's play proposes an alternative experiment. The latter, while taking into account the relationship with the Native environment, enables people from different cultures to engage in a fruitful dialogue. In her essay, Yvette Nolan concentrates on *Chasing Honey*, a 2007 play by Laura Shamas, a member of the Chickasaw nation. This powerful work focuses on the ancient practice of beekeeping as a marker of Native identity. The advent of Colony Collapse Disorder, as bees unexpectedly leave their hive, is likened to the decay of Native identity. It suggests a loss of interconnectedness with the natural world. In her contribution, Nicholle Dragone concentrates on *Re-Creation Story*, a 2008 play by Onondaga playwright Eric Gansworth. This work dramatizes the mythology of Haudenosaunee creation through a blend of Native storytelling and Western metatheatre. Indeed, the narrator of the play is invited to deliver a lecture about Haudenosaunee creation in a college-like setting. Only when he allows the audience to correct his narrative does Gansworth become a true Haudenosaunee storyteller. Thus, the play develops a Native ecology balancing the human and the non-human through a perpetual process of re-creation. In the concluding essay of the section on North American Indigenous drama, Ginny Ratsoy shows how Kevin Loring's first play, *Where the Blood Mixes* (2009), combines issues of ecospirituality and ecojustice. In this work, Loring foregrounds traditional N'lakap'mux (Salish Natives from interior British Columbia) concepts of interconnectedness of being and place. This spiritual environment can heal the wounds inflicted by the residential school system, which took generations of Native children away from their cultural roots. Not only is land of central importance in this work, but so are the Thompson and Fraser rivers, the habitat of sturgeons, the guardians of Native spirituality. The human and the non-human thus merge into a single web of being.

The second part of this volume, which is devoted to performance in Oceania, equally tackles issues of ecospirituality and ecojustice. Maryrose Casey's essay, "Serving the Living Land: Place and Belonging in Australian Aboriginal Dramas," analyzes the spiritual role of the land in works by contemporary artists Wesley Enoch and David Milroy. Casey successively examines the role of birthing trees in Enoch's *The Story of the Miracles at Cookie's Table*, the Aboriginal recontextualization of Euripidean models in Enoch's *Black Medea*, as well as the allegorical role of the vegetable garden in Milroy's *Windmill Baby*. While Casey focuses on text-based drama, Rachael Swain illustrates performance practice from remote Australia, in particular from the North-West. In "Dance, History and Country—An Uneasy Ecology in Australia," Swain develops an ecocritical performance mode which invites spectators to accept an Indigenous perception of "country," the Aboriginal English word meant to refer to the Australian outback. Drawing on her experience as a theatre practitioner, she carefully details how stories of the Dreaming are enacted in the performance, *Fire, Fire, Burning Bright*, devised by the Neminuwarlin Performance group (2002). She also focuses on *Crying Baby* (2000), by Marrugeku, an "intercultural company creating contemporary dance theatre in remote Indigenous communities" (172). She traces how Indigenous dance is used in this performance to tell the "Orphan dreaming" of an Aboriginal family forcibly removed from its "country" (172). In her essay, Diana Looser offers the first English-language examination of francophone Kanak playwright Pierre Gope's *La Parenthèse*. Gope, Looser argues, belongs to the growing number of Indigenous Pacific artists who have recently articulated ecological debates in the wake of colonization and globalization. In *La Parenthèse*, Gope uses the garden archetype to articulate a post-pastoral vision that reflects political issues relevant to contemporary New Caledonia. In this garden, the Gaiac tree symbolizes ancestral wisdom while Hortensia represents settler-colonial groups. The play asserts the profound link with the natural world characterizing Kanak culture. A cross-cultural vision of harmony is eventually established between different plants symbolizing ethnic groups. By way of conclusion, Looser underlines the originality of Gope's ecodramaturgy, which "mobilizes the natural environment to provide a more lucid view of the operation of human institutions" (197). She points out the critical role of the "mediation of the artist as a gardener" (197). In their essay, "Unfolding the Cloth: Patterns of Landscape and Identity in The Conch's *Masi*," Lisa Warrington and David O'Donnell analyze work devised by a Pacific theatre company operating from New Zealand. The Conch was founded by Nina Nawalowalo, a New Zealander of Fijian and English extraction. As the authors indicate, "*Masi* takes an ecological stance, embracing the inter-connectedness of art, land, identity, culture, politics, environment and spirituality" (199). As a performance piece, *Masi* blends Fijian and Western theatrical devices, including photographic illusion. Cast as an observer in the piece, Nina Nawalowalo introduces Fijian elements through her folding and unfolding of *masi* cloth. As the authors of the essay indicate, "*masi* cloth symbolizes the land and the natural world [...] the makers of *masi* cloth are predominantly female" (204). For Nawalowalo, *masi* cloth brings to mind the homeland as well as the soul of Fijian culture. It evokes the Polynesian notion of *va*, the space of "betweenness" that does not separate but rather connects (202). Furthermore, *Masi* develops a trickster-like relationship with its audiences, one that recalls Pacific satiric clowning traditions. In her contribution, Hilary Halba examines how the Māori concepts of "mauri (life-force) and tapu (restriction or protective prohibition, influence of ancestral entities)" (220) are articulated in the ecosystems of various Māori works: *Woman Far Walking* (a play by Witi Ihimaera), *When Sun and Moon Collide* (a play by Briar Grace-Smith) as well as in *The Strength of Water*, a film written by Briar Grace-Smith. In the latter work, the negative effects of the tapu of a particular place are alleviated by water. As Halba puts it, "a tapu is cleansed through dialogue between humans and the natural world" (231).

As a conclusion to the intercontinental journeys charted in the previous essays, Maufort's "Performing the Spirit of the Earth: Multi-faceted Aesthetics of Ecology in Contemporary Indigenous Drama" provides a comparative analysis of aesthetic devices in plays from North America and Oceania, focusing on works by Tomson Highway, Andrea James, and Hone Kouka. In doing so, it brings together the different regional ecosystems examined in this anthology. Echoing Maryann Henck's discussion of Drew Hayden Taylor's *The Berlin Blues* and Diana Looser's examination of *La Parenthèse* in the context of a globalized age, Maufort investigates how Aboriginal playwrights develop unique aesthetics enabling them to express their ecological vision in ways that evade the homogenizing gaze of Western globalization.

All in all, the essays assembled in this volume explore the multiple ways in which Indigenous playwrights from continents separated by the Pacific Ocean have negotiated their ecological interconnectedness with the natural world, whether in the shape of the land, rivers, animals, or plants. As a genre, Indigenous theatre seems particularly apt to spatially configure the relationships between humans and their environments: from the highly experimental approach to uranium mining in Marie Clements's *Burning Vision* to the ritual preparation of a dancing circle in James Luna's *Emendatio,* from the planetary, cosmopolitan vision of Tomson Highway's *Rose* to the metaphorical landscapes of Diane Glancy's plays, and from Wesley Enoch's dramatizations of Australian Aboriginal Dreamings to Hone Kouka's and Briar Grace-Smith's celebrations of the spiritual bond between Maori people and the land, the spectrum of "staging nature" is as wide as it is powerful. In dealing with ecospirituality or ecojustice concerns, the various performance practices discussed in this volume constitute, in their very diversity, a chorus of ecological voices. The songs of this chorus do not celebrate a facile harmony with nature but allow environmental challenges to be heard.

Works Cited

Adamson, Joni. *American Indian Literature, Environmental Justice, and Ecocriticism: The Middle Place.* Tucson: U of Arizona P, 2001.

— and Scott Slovic, eds. *Ethnicity and Ecocriticism.* Special Issue of *MELUS* 34.2 (Summer 2009).

Alaimo, Stacy. *Bodily Natures: Science, Environment, and the Material Self.* Bloomington: Indiana UP, 2010.

Arons, Wendy. "Introduction to Special Section on Performance and Ecology." *Theatre Topics* 17.2 (September 2007): 93-140.

— and Theresa J. May, eds. *Readings in Performance and Ecology.* New York: Palgrave Macmillan, 2012.

Bellarsi, Franca. "Ecocriticism, or Beyond Eurocentric and Anthropocentric Anti-Humanism." *Belgian Essays on Language and Literature* (2003): 155-66.

—. "The Challenges of Nature and Ecology." *Comparative American Studies* 7.2 (June 2009): 71-84.

Bhabha, Homi. *The Location of Culture.* London: Routledge, 1994.

Buell, Lawrence. *The Environmental Imagination: Thoreau, Nature Writing, and the Formation of American Culture.* Cambridge: Harvard UP, 1995.

—. *The Future of Environmental Criticism: Environmental Crisis and Literary Imagination.* Malden: Blackwell, 2005.

Cilano, Cara, and Elizabeth DeLoughrey. "Against Authenticity: Global Knowledges and Postcolonial Ecocriticism." *Interdisciplinary Studies in Literature and Environment* 14.1 (2007): 71-86.

Clark, Timothy. *The Cambridge Introduction to Literature and the Environment.* Cambridge: Cambridge UP, 2011.

DeLoughrey, Elizabeth, and George B. Handley, eds. *Postcolonial Ecologies: Literatures of the Environment.* Oxford: Oxford UP, 2011.

Egan, Gabriel. *Green Shakespeare: From Ecopolitics to Ecocriticism.* London: Routledge, 2006.

Garrard, Greg. *Ecocriticism.* 2nd ed. London: Routledge, 2012.

Glotfelty, Cheryll. "Introduction: Literary Studies in an Age of Environmental Crisis." Ed. Cheryll Glotfelty and Harold Fromm. *The Ecocriticism Reader: Landmarks in Literary Ecology.* Atlanta: U of Georgia P, 1996. xv-xxxvii.

Goodbody, Axel, and Kate Rigby, eds. *Ecocritical Theory: New European Approaches.* Charlottesville: U of Virginia P, 2011.

Gray, Nelson and Sheila Rabillard, eds. *Theatre in an Age of Eco-Crisis*. Special Issue of *Canadian Theatre Review* 144 (Fall 2010).

Heise, Ursula K. *Sense of Place and Sense of Planet: The Environmental Imagination of the Global.* Oxford: Oxford UP, 2008.

Huggan, Graham, and Helen Tiffin. *Postcolonial Ecocriticism: Literature, Animals, Environment.* London: Routledge, 2010.

Kershaw, Baz. *Theatre Ecology. Environments and Performance Events.* Cambridge: Cambridge UP, 2009.

Knowles, Ric, ed. *Interspecies Performance*. Special Issue of *Theatre Journal* 65.3 (October 2013).

Krech, Shepard E. *The Ecological Indian: Myth and History.* New York: Norton, 1999.

Naess, Arne. *Ecology, Community, and Lifestyle.* Cambridge: Cambridge UP, 1989.

Nelson, Robert. *Place and Vision: The Function of Landscape in Native American Fiction.* New York: Peter Lang, 1993.

Payne, Daniel G. *Voices in the Wilderness: American Nature Writing and Environmental Politics.* Hanover: UP of New England, 1996.

Richmond, Joy. "Introduction: Ecocriticism." *Journal of Dramatic Theory and Criticism* 20.2 (2006): 123-25.

Sessions, George. *Deep Ecology for the Twenty-First Century.* Boston: Shambhala, 1995.

Sinclair, Keith. *A History of New Zealand.* Auckland: Penguin Books, rev. ed. 2000.

Soja, Edward W. *Thirdspace: Journeys to Los Angeles and Other Real-and-Imagined Places.* Malden: Blackwell, 1996.

Ward, Russel. *Concise History of Australia.* St Lucia: U of Queensland P, 1965.

Zapf, Hubert. "The State of Ecocriticism and the Function of Literature as Cultural Ecology." Ed. Catrin Gersdorf and Sylvia Mayer. *Nature in Literary and Cultural Studies: Transatlantic Conversations on Ecocriticism.* Amsterdam: Rodopi, 2006. 49-70.

Stages of Resilience

Heteroholistic Environments in Plays by Marie Clements and Yvette Nolan

Birgit DÄWES

Universität Wien

I. (Not) Hugging Trees[1]

Between April and September 2013, the Kiasma Museum of Contemporary Art in Helsinki, Finland, exhibited a selection of works by Finnish artist Eija-Liisa Ahtila which deal, in the artist's words, with "the relationship between humanity, animals, and nature" ("Parallel Worlds"). In *Horizontal,* a moving image installation projected onto six separate screens (fig.1), Ahtila presents a "portrait" of a living spruce tree in its original size and setting ("Horizontal"), including the original on-site soundscape that was recorded when the tree was filmed. Since the tree is far too large to be reproduced in the limited exhibition room, the artist chooses a horizontal framework, as announced in the title. Yet, as she notes, even "[t]he horizontal film frame is unable to show the entire tree in one picture. If one uses a wide-angle lens instead, the tree appears distorted and is still too large to fit in one picture. Stepping back to allow the entire tree to appear inside the frame makes the picture no longer a portrait of a tree, but a landscape in which the tree is but one element" ("Horizontal, 2011"). Her solution is a combination of six separate close-up shots of parts of the tree, which are then projected together to give the impression of an organic whole.

1 I am very grateful to Yvette Nolan for her feedback on the draft version of this paper, and for kindly providing me with the manuscript of her play. Eija-Liisa Ahtila has kindly allowed me to use her work for this essay, for which I thank her profoundly. I would also like to thank my co-editor, Marc Maufort, for his untiring intellectual and practical support for this project.

Figure 1. Eija-Liisa Ahtila, Horizontal; photograph by Birgit Däwes, with kind permission from the artist.

While watching the film closely, however, viewers will soon notice that the six segments do not exactly fit: as the tree swerves gently in the breeze, the different parts and severed branches move against each other, sometimes in opposite directions, so that the impression of a unified whole is constantly disrupted. The authenticity of the setting, underlined by the fact that the artist includes herself in the first screen's upper left-hand corner for reasons of scale, is thus constructed as inherently elusive and unstable. This instability is further reinforced by the separation of the room into two sections, divided by a wide doorway. Parts of the projection are visible on the passageway below the ceiling; an effect which points once more to the impossibility of representing a natural tree in its original size, but which also highlights the mode and process of reproduction. By calling attention, within the spatial limitations of the mimetic approach, to the technology of successive projection as viewers walk through the installation, the work of art develops an aesthetic of critical awareness, combining the beauty of the tree and its soundscape with a materialized knowledge of the urban, institutional, museal space of its representation. The popular Western dichotomy of nature vs. culture is thus effectively perforated.

I would like to use this work of art as a figurative preface, or—in theatrical terminology—as a stage design for my approach to contemporary Indigenous North American drama. Using examples from Native American and First Nations plays, and with a special focus on Métis playwright Marie Clements's *The Edward Curtis Project* (2010) and Algonquin/Irish dramatist Yvette Nolan's *The Unplugging* (2012), my aim is to develop a thematic and methodological model—of what I call heteroholistic cultural ecology—that allows us to approach

Indigenous representations of nature and environment on stage, but that also goes beyond the notion of ecocriticism as simply concerned with environmentalism. Instead, by loosely following Hubert Zapf, I propose to read the semiotics of theater or other art forms themselves as cultural ecosystems.

Whereas ecocriticism has, at least in part over the past decade, been aiming at more inclusive understandings of the relationships between humans and their environment, critics have only very recently begun to explore the intersections between Indigenous cultures, ecological issues, and their cultural representations.[2] One of the major impediments of a differentiated ecocritical approach to Indigenous cultures is the strong and tenacious presence of the "eco-Indian," a Romantic stereotype fueled, for instance, by the *Keep America Beautiful* campaigns of the 1970s and 1980s, which show Iron-Eyes Cody[3] with a feather in his hair, crying at the pollution of the environment. As an escapist and reductive fantasy, the idea of the "Native North American as ecologist and conservationist" (Krech 16), who lives in perfect harmony with nature, perpetuates the colonial agenda of displacing Indigenous people to the past; coding them by noble simplicity and uncivilized life and thus unfit for the challenges of contemporary society. Just as Ahtila's tree dissolves the illusion of a harmonious, holistic nature by splitting the tree into six segments, many contemporary works of First Nations art expose the ideological trappings of the "eco-Indian" and replace them by differentiated and more complex engagements with nature. "The attempt to portray a spruce tree," Ahtila asserts, "[...] invites us to consider the preconditions of anthropomorphic dramaturgy and the resultant valuations in images

2 Even though Lawrence Buell and other ecocritics before him refer to Native American texts in their explorations of literature and the environment, a first systematic approach was undertaken by Joni Adamson in her study *American Indian Literature, Environmental Justice, and Ecocriticism: The Middle Place* (2001), which asserts the necessity "to theorize a way of reading that provides us with the tools we need for building a more satisfying multicultural ecocriticism and a more inclusive, multicultural environmentalism" (185). While Adamson's reading is deeply rooted in the activist environmental justice movement (cf. Adamson), Robert Nelson and Lee Schweninger expand the theoretical framework by studies on Indigenous literatures and their narrative constructions of landscape (Nelson, Schweninger). In a study on *Ecocriticism: Creating Self and Place in Environmental and American Indian Literatures* (2002), Donelle Dreese focuses even more closely on the precise interconnections between Self and Place by examining Native American poetry and prose, yet her book is of limited scope (covering only 116 pages) and thus marks but a first stepping stone toward further research. See also the Introduction to this volume for a more detailed survey.

3 Espera Oscar de Corti, or "Iron Eyes Cody" was an Italian-American actor who frequently played Native American roles and claimed to have Cherokee ancestry.

and the order of presentation" ("Horizontal, 2011"). It is precisely this "anthropomorphic dramaturgy" that many Native North American plays address; and like the Finnish artist, they develop various counterdiscourses, or counter-ideological choreographies, to design environments on stage.

Furthermore, in addition to dissolving the boundary between culture and nature, and to setting common patterns of perception on end, Eija-Liisa Ahtila's installation can be read as an encouragement to approach the methodology of ecocriticism itself horizontally, or sideways, for a fresh perspective on the intersections between place, environment, and cultural representation. Whereas much ecocritical work today either strongly prioritizes systemic engagements with nature at the expense of questions of cultural difference and specific historicities, or finds itself driven by an environmentalist agenda that tends to neglect issues of representation and aesthetics, my "sideways" approach combines two methodological concepts that will help us to re-examine patterns of human-environmental interactions within a Native context, and to create a more abstract framework for the analysis of representations of nature on stage. These two concepts are cultural ecology—a term within which Hubert Zapf frames literature "as a specifically powerful form of symbolically staging and exploring the interrelationship between cultural and natural processes and energies" (49)—and heteroholism, which was recently developed as a new reading strategy for Native American literatures by Billy J. Stratton. Heteroholism, according to Stratton, marks "the interplay between intellectual and spiritual or visionary ways of knowing that are anchored in specific spatial and temporal sites that distinguish Native worldviews and knowledge" (148). It acknowledges and applauds the culturally specific aspects of Indigenous works of art while refusing to reduce them to essentialist or narrowly demarcated parameters of description.

Reflected through these theoretical lenses, Marie Clements's and Yvette Nolan's plays present intriguingly innovative representations of nature on stage: they effectively dodge the paralytic discourse of the eco-Indian stereotype; they disrupt the Romantic notions of nature as a harmonious environment; they dissolve any dichotomous logic of nature-versus-culture; and, by designing culturally and historically specific stages for Indigenous codings of land and environment, they also effectively share an anti-colonial political agenda. In line with Eija-Liisa Ahtila's *Horizontal,* this approach to First Nations cultural ecologies on stage is not about hugging trees at all, but, on the contrary, about dissecting and examining their projections—in order to find a more differentiated theoretical ground for the analysis of Indigenous American representations of nature. Furthermore, the plays' specific engagements with space and land, as well as with time and temporality as crucial constituents of their encoded environments, constitute theatrical ecosystems in their own right; they not only contribute to the richly diverse fields of Indigenous Studies, Canadian Studies and Theater Studies, but they help us to reconsider, from new and unexpected angles, the frameworks and methodologies of ecocriticism as a reading practice.

II. Reading Environments: Indigenous Cultural Ecologies

Nature, power, and cultural identity are inseparably intertwined; yet contemporary ecological discourse tends to blur these interconnections in favor of politically more efficient imaginaries of harmony or beauty. As Stephanie Rutherford has recently criticized, much of popular environmentalism is actually a restrictive form of governing: at specific cultural sites where our perception of nature is defined and affirmed (such as museums of natural history or national parks), human beings are invited to align their identities with the power structures of late capitalism, becoming "the environmental citizen as consuming subject" (Rutherford 194). In the field of Native North American Studies, the reductive effect of such commodifying moves is particularly harmful. On the one hand, most Indigenous North American cultures emphasize the interconnectedness of all beings and thus emblematize what Timothy Morton terms "the ecological thought"—the ability "to join the dots and see that everything is interconnected" (*Ecological* 1). Choctaw historian Clara Sue Kidwell points out that the major difference between European and Native American systems of knowledge lies in the "role of human beings in the processes of nature": In Indigenous cosmologies, "[h]uman beings are an integral part of the [...] natural world, not observers who stand outside these processes" (98). On the other hand, as Shepard Krech III reminds us, the image of the eco-Indian, to which this difference is often reduced, most problematically impacts upon political and cultural sovereignty:

> It masks cultural diversity. It occludes its actual connection to the behavior it purports to explain. Moreover, because it has entered the realm of common sense and as received wisdom is perceived as fundamental truth, it serves to deflect any desire to fathom or confront the evidence for relationships between Indians and the environment. (27)

These relationships have, in fact, not been harmonious at all: "Native Americans have not always been environmental perfectionists. Occasionally, in pre-Columbian times, native urban areas taxed the local environment to the point of self-destruction (e.g., by overgrazing and razing of forests)" (Grinde and Johansen 11). Furthermore, as Ahtila's *Horizontal* also nicely illustrates, harmonious relationships are not even the point: on the contrary, Ursula Heise emphasizes that recent scientific studies in ecology substantially challenge the conventional notions of 'harmony' and systemic stability in nature as such: contemporary biological research much rather presents "a more complex image of ecosystems as dynamic, perpetually changing, and often far from stable or balanced" ("Hitchhiker's Guide" 510).

In recognizing the need for such "a more complex image," I am suggesting here to use the concept of cultural ecology as a basis for the critical inquiry into Native North American representations of nature and environment. Whereas the first, ethnology-based definitions of cultural ecology focused on "ascertaining how the adaptation of a culture to its environment may entail certain changes" and on "determin[ing] whether similar adjustments occur in similar environments" (Steward *Theory of Culture Change* 9; see also Frake), more recent approaches—initiated by Gregory Bateson and further developed by William Paulson—understand culture itself as an eco-system, whose elements and participants are tied together by fragile, interdependent relations. Hubert Zapf has added a functional model of literature to these meanings: Cultural ecology, in Zapf's sense, is thus a method that considers literature and other modes of artistic expression as part of an encompassing cultural ecosystem, functioning "both as a sensorium for the deficits and imbalances of the larger culture, and as the site of a constant renewal of cultural creativity" (Zapf 49). "Literature," he elaborates (and we may well add theater and performance),

> appears as a *sensorium and imaginative sounding board* for hidden problems, deficits, and imbalances of the larger culture, as a form of textuality which critically balances and symbolically articulates what is marginalized, neglected, repressed or excluded by dominant historical power structures, systems of discourse, and forms of life. (56 [emphasis in the original])

Departing from Zapf's seminal expansion of the concept—not simply by analogy, but by a functional revision—the use of cultural ecology as a viable and inclusive methodology to explore human interactions with their environments may help us to navigate around the impasses of dualistic thinking (nature vs. culture/civilization, conservationist vs. progressive) and reductive equations. Cultural ecology promotes the critical awareness that space, place, and landscape are always cultural products, designed and encoded within specific economies and political contexts for specific purposes; and it acknowledges the importance of cultural difference in the mapping of space, time, and community. More specifically, cultural ecology serves as a framework to understand the ways in which culturally specific narratives (in this case, by Indigenous playwrights) creatively interact with their North American environments: not only textually and performatively representing their coordinates (of space, time, land, territory, cosmos, or nature), but dynamically (re)constructing them; seismographically registering imbalances within their cultural systems, and counteracting upon them.

With regard to contemporary First Nations plays, cultural ecology allows us to explore quite specifically, and in differentiated ways, the performative strategies by which Indigenous artists design space, time, and landscape on stage, and how they map them into empowered environments for the negotiation of cultural identities—both locally and on a global scale. In general anthropological terms, human beings define themselves in relation to their environments, situating themselves within spatial and temporal settings, as well as by coordinates of social, cultural, and—to a larger or lesser extent—religious and metaphysical surroundings. "Identity and environment," in Nathan Straight's words, "are mutually constitutive" (12). In addition to a differentiated analysis of the actual stagings of Native American environments (including their semiotics of 'land' and 'nature,' the interrelatedness of their elements, as well as their non-linear patterns of time), the mapping of an Indigenous cultural ecology therefore requires a differentiated look at the configuration and performance of identity—both individual and collective.

Joseph Bauerkemper deplores that critical engagements with Native North American literature often diagnose these texts' nonlinear "histories and chronologies," but "beyond underscoring their presence as markers of cultural-groundedness, the exploration of the social and political significance of nonlinear histories and chronologies in American Indian literatures remains neglected" (28). The impact of these differences on Indigenous identities is crucial, not only for the acknowledgement of the social and political implications of space and time, but for a larger understanding of Native environments and Native cosmologies at large. Over the past two decades, quite a number of theoretical approaches have been undertaken to develop terminologies for such cosmological and cultural differences, from Jace Weaver's "communitism"—the acknowledgement that "Native literatures differ from dominant discourse in their commitment to community" (163), which "includes a shared sense of story," and an understanding of nature that is marked by a "potent sense of interrelatedness," among others (164)—to Gerald Vizenor's "postindian warrior" stance and Floyd Favel's theater-based model of "Tradition ←→ Process ←→ Performance" (see Mojica 129-31). Cherokee scholar Billy J. Stratton adds to these an expansion of what he terms, with reference to a concept by Tom Holm *et al.*, "the Peoplehood matrix" (148), which "considers the status of individual Native cultures based upon four interconnected cultural elements that all communities share: language, sacred history, territory, and ceremonial practice" (148). These four constituents can be summarized in a "heteroholistic ordering of the world" (148), which acknowledges the "intimate spiritual and cultural relationships to the land" on which Native American communities rely. The importance of local knowledges is central to the concept; at the same time, however, it is "conceived as heteroholistic because it emerges from specific geographic locations that are universally valid, while it simultaneously coexists alongside other universal, but localized, tribal experiences [...]" (Stratton 149). Heteroholism, as a complementary theoretical matrix to cultural ecology, acknowledges the importance of land and local landscape to Indigenous cultures without denying the dynamic, in-progress, and unstable characteristics of these cultures. Together with the concept of heteroholism, I argue here, cultural ecology prevents us from falling into the multiple traps of resorting to what Kate Soper terms "uncritical ecological naturalism" (149-50), of conflating Native ecologies with the notion of the "eco-Indian," of confusing representations of nature and land with ethnographic evidence, or of reducing environmental knowledge to culturally insensitive universalisms.

III. Indigenous Ecologies on Stage

As in all forms of Native North American cultural expression, the land features centrally in performance and theater. Robert Shorty's play *Na Haaz Zan* (1969), for instance, begins with a specific location, envisioned for the audience by the narrator: "A long time ago when time first began [...] the first land they call it" (n. pag.). This place provides the specific setting for the play, but it also functions as the symbolic location of the Navajo people's collective origin, initiating, framing, and explaining their collective identity. In Elizabeth Woody's words: "[l]and is not exclusive property but is the embodiment of our ancestors" (165). The physical spaces underlying cultural ecologies are thus never merely geographical or material; from an Indigenous North American perspective, they shape and sustain human identity as such. "[I]t is very clear that the landscape is part of the people, or that they are the landscape, or the landscape is a map of their own visions and experience," Anna Lee Walters writes in *Talking Indian* (104): even as a source of identity, the land is never merely the basis for political claims to sovereignty, but it provides natural and cultural resources and a sense of home as the inclusive marker of social, spiritual, and mythical coherence.

There are various ways of translating this comprehensive, or heteroholistic, understanding of the land into a theatrical dimension. The rather obvious solutions include outlines of set, stage decoration and the construction of landscape through dialogue and character movements.

In Hanay Geiogamah's *49* (1975), the setting is "a ceremonial ground circa 1885 and the same ceremonial ground in the present" (85), which is marked by "brush arbors in the background, and a roadway extend[ing] out of the area" (85). In this setting, the plot unfolds: at a 49—the nighttime ceremonials that follow "the more formal activities of the powwow" (87)—a group of adolescent Native Americans gathers in order to sing, dance, and share stories. They are surrounded by the threatening presence of the institutionalized white forces, personified by the highway patrol. Only audible through their police radios, the officers decide to move in on the meeting, but they find themselves confronted with "an elaborate barricade" (129) of some two thousand (111) 49ers who refuse to leave. At the ceremonial grounds on which the action takes place, the group is guided by the "Night Walker," an ageless ceremonial leader who moves between 1885 and 1970 to provide traditional knowledge and communal coherence.

The spaces between the Indigenous singers and dancers on the one hand, and the police patrol on the other, are marked visually and audibly by the use of music, sound, and light: the chanting and drumming of the group, for instance, is juxtaposed with the "sudden, jagged cackle of a two-way police radio" (92). In similar expressionistic style, the light is used to charge the ceremonial, ancient land with additional meaning: In scene three, for instance, Night Walker's ritualistic movement through "rugged terrain" is accompanied by "odd flashes of light" (95); the "harsh red lights" of scene six mark the historical time shift to 1885 (103); and the "flashes of patrol car spotlights [which] pierce the dark" in scene eleven indicate the approaching threat of the police. As the setting is developed into a cultural ecology of interrelated elements, one central signifier emerges as a symbol of the connections between past and present, material and mythical sites, and between traditional and contemporary orders of knowledge: the arbor at the ceremonial ground, which combines notions of the land with the people's cultural heritage. "This arbor cannot be killed," Night Walker reiterates. "It is strong and powerful. It has lived for a very long time. It can be burned and torn apart, but its life cannot be taken from it" (122). He validates this comprehensive meaning by telling a story, uniting past and present, and appealing to the young people to use this arbor as a site of spiritual and cultural revival: "You will pray at the arbor for many seasons. Pray for our tribe to gather beneath it" (125). At the end of the scene, "[t]he arbor comes blazingly alight" (125) in a literally spotlighted manifestation of the land's spiritual and transcendental power, as expressed by the verbal and visual construction of this set.

The "first land" in *Na Haaz Zan* and the arbor in *49* nicely illustrate the concept of a heteroholistic cultural ecology: firmly grounded in local sites, but trans-tribally extending their meanings, they represent the comprehensive, interwoven dimensions that shape physical spaces in Native theater. These dimensions never nostalgically reduce "nature" to a Romantic backdrop or projection screen, but they are always socially and culturally encoded, weaving coordinates of space, time, storytelling, and communal identity into the specific locations of the homeland. Or, as Jaye Darby puts it, "[f]ar more than just a geographic location, home is the timeless source of spirituality, the site where intergenerational continuity and community nurture the continuance of traditions and sacred responsibilities" (55).

IV. Marie Clements's *The Edward Curtis Project*

Marie Clements's play *The Edward Curtis Project,* first produced at the Presentation House Theater in Vancouver in 2010, begins on precisely this note, emphasizing the embeddedness of individuals within traditions and communities across time, the confluence of which is microcosmically focused and rendered visible at specific spatial sites. The cultural ecology of *The Edward Curtis Project*, I argue here, is thus structured into a threefold landscape—of space, time, and identity—whose layers are dynamic, interdependent, and inseparably intertwined from the very first scene. The play opens with the protagonist, Angeline, on an empty stage with family photographs projected all around her. A contemporary Dene Saulteaux journalist, Angeline just accepted an award for a story she wrote about three First Nations children "frozen to death" in the Arctic (Clements 12). The placement of Angeline among projections of her ancestors and family embeds the theatrical space in an environment of tradition, family, and community, while the initial stage direction also emphasizes the changing and fluid character of both spatial and temporal dimensions: The set is described as "[e]xtremely bare except for walls of light and dark that are able to transform space and time" (9). As Angeline's sister Clara enters to give her one of the volumes of Edward Curtis's *The North American Indian*, this transformation is made visible: "*Edward Curtis begins to speak*" (16), and the characters cross the boundary between 1911 and 2008 to meet and interact. In line with the heteroholistic notion of permeable spaces and time levels, the situations of the early 1900s are merged with the present throughout the play; moving from the Pacific Northwest to the Prairies, and from the Arctic to New Mexico, among others.

The contemporary and historical identities arising from these fluid environments are accordingly coded as relational, fragmented, decentered, and in-progress: the nine characters are impersonated by merely four actors. The play circles outward from the figure of Edward

Curtis, setting him in dialogue with other historical characters: his wife Clara, his Crow interpreter Alexander Upshaw, and a Duwamish woman he named "Princess Angeline"—one of the first Native people he photographed, and who posed for him repeatedly. Illustrating Susan Bennett's claim that on the theatrical stage, the live, performing body functions as an "archive, the literal vessel of a somatic history" (35), these characters shift back and forth into contemporary Native characters of the same names; centering on Dené journalist Angeline, her sister Clara, as well as her boyfriend Yiska—who physically merges with Upshaw.[4] The play thus resists any simplifying dichotomies along ethnic, generational, or economic lines of difference, since the boundaries between Angeline, Curtis, and Clara are frequently blurred in shared characteristics. The contemporary Angeline is linked to Curtis by her profession as a journalist; Angeline is depressed, just as Curtis's wife Clara is; and her sister, like Clara Curtis, has to go through a divorce. Individual identities cannot be disentangled from their contextual communities, but they are—as in an ecological system—defined by their interdependencies with the Other. This effect is further underscored by the play's accompanying exhibition of over one hundred photographs taken by Rita Leistner, a non-Native Canadian photojournalist. Composed in diptychs that overlap past and present identities—through traditional costumes such as Haida blankets, for instance, in contrast to fashionable contemporary rain boots—these photographs "create a visual dialogue or bridge between past and present, between the legacy of Edward Curtis and the colonial past, and the bright future elucidated by many of the people I met in my travels" (Leistner 73).

In addition to arranging the stage and characters in a dynamic and constantly shifting system of interrelations complemented by contemporary images, *The Edward Curtis Project*'s cultural ecology plays out on a thematic and symbolic level. The central image, developed from Curtis's photographic agenda, is that of the "vanishing race," as a famous 1904 picture in volume 1 of *The North American Indian* was entitled. Capturing a popular topos of the time, the notion of the "vanishing

[4] As a respected "advocate and campaigner for tribal rights" (Zamir 615), the historical Alexander Upshaw negotiated on Curtis's behalf and provided the photographer with access to Native people and sacred traditions. He also graduated from Carlisle Indian School in 1897, worked as a teacher, and was married to a non-Native woman in Nebraska. In Curtis's portfolio, however, Upshaw is presented in the dress of a Crow warrior—catering to a market that wanted to see "traditional," noble-savage based images. Analyzing this image, Shamoon Zamir notes that it is unlikely the headdress and necklace were actually Upshaw's, since that specific type of clothing was worn by leaders or elders (cf. 630). Zamir argues that it is precisely this "masked quality" and "the very theatricality of the image" which indicates Upshaw's agency: "his performance was the vehicle for a genuine self-expression" (648).

(Native) American" frames a history of genocide, dispossession, and imperialist disempowerment in the terminology of a natural process—as if Indigenous Americans had simply become extinct on their own account. Ecological extinction, according to biologists, is defined as "the reduction of a species to such low abundance that, although it is still present in the community, it no longer interacts significantly with other species" (Estes *et al.* 253): the ideology of the "vanishing race," accordingly, dissolves any blame for this situation in the rhetoric of a natural given. The politics of systematic annihilation, racism, treaty violations, forced assimilation, and economic injustice are covered up in the soothing message that Nature has terminally solved any conflicts. While historically the argument of decreasing population numbers was not inaccurate,[5] its instrumentalization within or even *as* an ecological process sealed the issues of power and ownership with the finality of Providence.

Marie Clements's play disrupts this theme in various ways: by dissecting and rewriting its origins, by reframing it within the contemporary scene, and by replacing it with an alternative discourse of continuity, futurity, and survival. "What happens," Rita Leistner asks in the preface to her exhibition, "when the 'vanishing race' doesn't vanish? What is the impact on a people who are told they are vanishing?" (71). Clements answers this question by a direct confrontation across the twentieth century. While the historical character of Edward Curtis is presenting his pictures to a large audience in New York's Carnegie Hall (29), the contemporary Angeline and her Salish boyfriend Yiska actively subvert the show. The stage direction stipulates that "*Edward Curtis photographs project on ANGELINE and YISKA, who begin to animate them. They look at EDWARD CURTIS throughout*" (47). Angeline—whose performing body is the living archive of the historical Princess Angeline—literally stops the simulation: Curtis notices that something is wrong and has to end the show. In the larger cultural ecosystem of contradictory signifiers, therefore, the ecologically disguised narrative of the "vanishing American" is overwritten by a living Native presence. The contemporary characters also literally talk back to Curtis and question his project: Whereas Curtis emphasizes the Natives' gratefulness for "preserving [their culture] forever" (28), Angeline and Yiska respond that "[y]ou were taking pictures of your idea of them. Big difference." (53). Curtis's authoritative declaration that the photographs are realities—which corresponds to the ideology of a 'natural' process of extinction—is dismantled by Angeline's questions, which emphasize

5 Russell Thornton calculates that the estimated five or six million aboriginal people who lived in the area of the present United States upon European arrival were decimated to a number of 250,000 around 1900 (133).

the cultural constructedness: "Are they? Or are they just perceptions? And if so, of whom? Those who take the picture or those who pose for them?" (28). The play here also questions, more or less implicitly, the representability of nature as such: Like Ahtila's spruce tree, which is reproduced as it was filmed, but arranged horizontally and in fragments, *The Edward Curtis Project* mixes authentic historical dates and materials with strategies that point to the constructedness of all history (or of nature and natural processes). Many of the words that Curtis speaks in the play are taken from the record of his 1911 show in New York, and the play also uses the music originally composed by Henry Gilbert for Curtis's show. Clements thus combines the historical sources from the past with alternative perspectives and contingencies to map a larger cultural ecology of the present and future.

Finally, in a very specific sense, Clements's play uses an Arctic scenery as a site for mapping the coordinates of Indigenous ecology, economy, and representation. When Angeline is first introduced, she remembers a story she published, for which she was given an award. As she starts telling the story, however, her voice is drowned out by the sound of the newscast:

> This just in from the far north...Three Aboriginal children were found frozen to death in the snow as temperatures plummeted to the minus-forties and their father, age 24, was found drunk. Treated for frostbite and hypothermia, the children's father said his youngest daughter was sick, so he had ventured out to get help and got lost in an icy field...The father told his sister at some point...he dropped the kids... (12)

The story is based on an actual incident from January 2008: a man named Christopher Pauchay left two of his children in the snow on Saskatchewan's Yellow Quill Reserve (cf. Couture 14). In the play, following the advice of her editor to mention only the basic facts, Angeline turns the story into an award-winning newspaper report. Yet instead of "writ[ing] the whole story" (45), she finds herself guilty of abbreviating the narrative into an individual tragedy of alcohol abuse. As a consequence, her narrative of the children's deaths persistently haunts her, causing the depression that she suffers from throughout the play. By blaming the Aboriginal father in her story, she tells Clara and Yiska, she became guilty of silencing the sociopolitical and economic contexts of the case, the systemic violence in which it was embedded, and thus its larger cultural ecology:

> I should have written that the father of those children was so young, so poor... living in a house that was so contaminated it should have been torn down... living between cardboard walls with no food, no clean water, no phone, no heat, and the only reason he decided to go out into minus-thirty-eight weather was because one of his kids was sick ... He went out to get help ... Do you think it was all his fault? Or maybe we all should own a little piece of it? (46)

Just like Curtis's portraits—which depict the Native subjects as "frozen in time" (65)—Angeline's report naturalizes the deaths into an image of vanishing; erasing, in its snippet view, the larger historical narratives of cause and effect. At the end of the play, however, this story is re-enacted in a collective effort, in which the environment's hostility is played out on geographical and climatic, as well as social and political terrains. In addition to the snowstorm, ice, and (literal as well as symbolic) impressions of "whiteness" (63) that shape Arctic ecology and forbid any romantic involvement with nature, Clements adds the character of the Hunger Chief, an impersonation of the timeless economic forces of poverty and injustice that dominate the Indigenous environment.[6] As the Hunger Chief confronts Edward Curtis, charging him with exploiting the people he photographed, he repeatedly asserts that "I want to be warm. I want to be well. I want to live" (62). Through Angeline's simultaneous presence and evocation of the frozen children, the Hunger Chief turns into a voice of the father silenced by Angeline—a voice that is representative of Aboriginal Canadians in general.

As the Hunger Chief exits, Edward Curtis, Dr. Clara, Angeline and Yiska take turns in telling a more detailed version of the story, in which Angeline allows herself to feel the full force of her emotions as she saw the children—a baby, a four-year-old, and an eight-year-old—frozen to death and "so alone" (65). Their collaborative re-enactment of empathy allows for a diversified perspective on the case, including the systemic coordinates of colonialism, poverty, and social injustice. When Angeline tells Curtis to take a picture of the frozen children, he rejects the responsibility "because I am not alive. You are" (66)—and begins to fade, accompanied by a caption reading "Edward Curtis, 1868-1952" (66). The vanishing photographer visualizes a reversal of the ethnographic hierarchy, of course, but he also points to the responsibility that comes with survival, on which the play ends. Angeline ultimately accepts her co-authorship of the narrative and ends the play on a note of hope and connection: "We have survived across time," she tells Yiska, "across place, to love each other towards a new day" (67).

In the cultural ecology of Marie Clements's play, the discourse of vanishing is powerfully turned around and relativized across different times and spaces. Far from promoting eco-Indian imagery, the play presents nature as a harsh and hostile environment, overlapped by and interdependent with the historical heritage of colonialism, and a social fabric of capitalist exploitation and injustice. On stage, this scenery is

---

6 In Rita Leistner's pictures, this emphasis is echoed by pictures of First Nations people living in appalling poverty on Vancouver's Downtown Eastside (94-95, 106-09).

complemented by "ways of knowing that are anchored in specific spatial and temporal sites that distinguish Native worldviews and knowledge" (Stratton 148)—which transcend boundaries of historical periods, languages, or ethnic groups in a heteroholistic move.

This principle of border-crossing, finally, is also underlined on the structural level: instead of acts or scenes, twenty-four "frame shifts" connect the different characters and time levels in smooth transitions. The frame shifts, as an image originating in the visual arts, defy any claim of representing nature or history mimetically; instead, the spatial or temporal environments that are presented on stage have already been encoded, selected, shaped, and constructed by the lens of a camera. Or, in more theoretical terms, in the cultural-ecological system of *The Edward Curtis Project*, which is itself part of larger cultural ecologies (of First Nations theater, or of contemporary Canadian art in general), nature is always already culturally encoded—a crucial methodological statement on ecocriticism as such.

V. Yvette Nolan's *The Unplugging*

Whereas the snow in Marie Clements's *Edward Curtis Project* connects contemporary characters to the injustices of the past, the icy landscape of Yvette Nolan's play *The Unplugging* is the site of a microcosmic exploration of the future.[7] Fully produced for the first time at the Arts Club Theatre in Vancouver in October 2012, *The Unplugging* also develops a heteroholistic cultural ecology from the layers of space, time, and identity. The play is loosely based on Velma Wallis's 1993 novel *Two Old Women*, a traditional tale of a nomadic Athabascan community, the Gwich'in, during a severe winter famine. In Wallis's novel, the community decides to abandon two of their elders, an eighty-year-old and a seventy-five-year-old woman. Hoping to save resources, they leave the two old women behind, much as in Jack London's 1901 story "The Law of Life," in which a blind old Native man is abandoned and killed by wolves. Wallis's tale, however, takes a different turn: instead of resigning themselves to their fate, the two women decide to fight for their survival. Establishing a campsite of their own, they use their knowledge and skills to build a snow shelter, collect firewood, and trap rabbits. Not only do they survive the winter, but through an extremely efficient resource management, they are able to gather

7 We learn in the first scene that the action takes place "fifty years" after Robert Frost's death, which was in 1963. Since the play was first produced in 2012, the temporal position just one year into the future gives the setting enough futurity to underline the play's utopian outline, but without losing its proximity to the present day in distant or completely unfamiliar science-fiction-like scenarios.

a generous surplus of food and clothing during the following spring and summer. The women's survival is particularly remarkable in light of the various hazards they have to face. In addition to struggling for minimum amounts of nourishment and against exposure, the women are continuously threatened by "the savage wolves that howled in the distance" (Wallis 67) and by fellow human beings: "They also were afraid that potential enemies might come upon them. Other bands were traveling, too, even in the cold winter, and the women did not want to be exposed to such dangers" (Wallis 33-34). And even worse than the threat from enemies, they struggle with the "broken trust" toward their own tribe and families: They knew "that the cold weather would force people to do desperate things to survive, remembering the taboo stories handed down for generations about how some had turned into cannibals to survive" (Wallis 34). The protective power of the community, this tale highlights, can just as easily be reversed. Yet the novel has an optimistic outlook: when the community eventually comes looking for the abandoned women one year later, they realize their moral mistake and agree upon a didactic "new beginning": "The People kept their promise. They never again abandoned any elder. They had learned a lesson taught by two whom they came to love and care for until each died a truly happy old woman" (Wallis 136). This ending crucially highlights the inseparability of ecology from social frameworks and ethical responsibilities: even in the most hostile of environments, communal solidarity and mutual respect are as essential for human survival as traditional ecological skills.

Yvette Nolan's play uses the same plot pattern and setting—in a cold Northern landscape between December and May—underlining once more the hazards of a naïve or romantic view on nature. As in Wallis's novel, too, two women, Bernadette and Elena, struggle for their survival by hunting rabbits and moose, and by battling the depression of being expelled from their community. Through its detailed and unadorned depiction of what James Ruppert calls a "subsistence lifestyle" (671), the play thus disrupts any Western nostalgic desires for natural harmony, effectively de-romanticizing any connection between Indigenous identity and nature. At the same time, Nolan's play also modifies Velma Wallis's *Two Old Women* in four crucial ways. First of all, the play is set in a decidedly contemporary environment, with references to Google (7), Disney (14), video games and Twitter (28), tying audiences much more closely to the diegetic world of the play. Second, the exiled protagonists are in their fifties and thus much younger than in Wallis's novel. While Bernadette jokes that "[f]ifty is the new eighty, I guess" (4), this difference highlights the obsession with youthfulness in contemporary Western societies and underlines its absurdity all the more plainly, given that these capable and strong women were banished from the community because they were considered "old" (4) and thus "found wanting" (14). Third, in addition to the two women, a third, non-Indigenous character adds diversity and dramatic energy to the plot: while Bernadette and Elena establish a new home at a log cabin, a younger man intrudes and alters their friendship. When Seamus, an urban bike courier and the son of Irish immigrants, who claims he was also banished by the community, seeks shelter and food, Bernadette and Elena have different opinions on how to deal with him, which reflect two different survival ethics. Bernadette, who misses intimacy, trades moose meat for news from the community and eventually gets sexually involved with him, whereas Elena does not trust him on accounts of gender: "Don't believe him. Bernadette. He's a man" (26). Finally, the fourth and most obvious difference to Wallis's text is the nature and extent of the underlying crisis. While the conflict in *Two Old Women* is triggered by a regular, locally restricted winter famine, Yvette Nolan's characters have survived an apocalyptic event on a global scale. After the title's "unplugging," an unspecified event that "destroy[ed] half the world" (19), leaving humanity without technology or electric power, societies fell into a state of "looting and carnage" (13) before new forms of authoritative leadership replaced the previous democratic structures. Nothing in this world is as it used to be: all kinds of information and knowledge "disappeared in a blink" (7), and capitalism was evidently ended when "money bec[ame] nothing more than a firestarter" (14).

For ecocritics, this play is interesting in obvious ways, since both the reason for the apocalypse and its outcome carry an environmentalist message: the "unplugging" is interpreted by Bernadette as a consequence of human ignorance toward sustainability. "This is how we got into this situation in the first place," she says, "people indiscriminately taking and using and wasting without any thought to the future" (21), which is why Bernadette thinks of the violent disruption "as the earth waking and shaking like some great dog, and all the machines and wires being shaken off like so many fleas" (19). By contrast, the deserted cabin in the woods they find is "built like a temple to nature" (12), with "[o]pen spaces, windows that let the sun shine in, simple, elegant, functional" (12). In a world of careless exploitation, the play tells us, a violent ending is a necessary consequence. This process of what biologists call an "ecological collapse"—the "rapid decline of ecosystem capacity [...] or disintegration of an ecosystem" (Curren 337), however, is not an end or message in itself in Yvette Nolan's play, and it should not be read as such. Instead, the crash, functioning as a catalyst for larger shifts in a global system, serves as a lens for the examination of social, political, and cultural ecologies. In scene four, we learn that Bern and Elena were banished from their community because their social system had changed completely: after the apocalypse, people began to “line up behind feudal lords” (14) in new communal structures. An individual leader named Laird took over, decided that only “women of child-bearing age” would be useful (15) and thus reduced all females to “breed-mares” (15). Elena mentions the violence that broke out: “You kept us alive. You kept me alive. I would’ve hung around like a fort Indian until I starved or Laird’s goons shot me, the way they did the feral dogs” (15).

Her mention of the “goons” here, in connection with the “fort Indian,” is important, since it evokes associations with the political situation on the Pine Ridge Reservation, South Dakota, in the early 1970s, when Oglala Lakota tribal president Richard “Dick” Wilson was charged with corruption and threatened with impeachment. For his own protection, Wilson had established the “Guardians of the Oglala Nation,” or GOONs, a paramilitary group of family members and friends of Wilson’s that intimidated and oppressed Wilson’s opponents (cf. Matthiessen 60-63; 131-32). The ensuing protest led to the siege of Wounded Knee in 1973, which was followed by a period of extreme violence, with many murder cases unsolved to the present day. In *The Unplugging*, the community’s situation also rapidly deteriorates after the expulsion of the elders: “it’s not going so well there,” Seamus tells the two women, claiming he himself was banished too. “There’s lots of—sickness, lots of anger, lots of power stuff going on. Laird is, like, king or something. He said I had to go because I wasn’t pulling my weight, but I think it was—something else” (28). This shift in the social ecology leads to his escape—or at least what we and the women are led to believe is an escape—from the oppressive environment.

Elena and Bernadette have two different strategies of dealing with this shift in their own newly established system: while Bern feels obliged to help Seamus, wanting to include him in their social setup, Elena is more skeptical and believes the intruder has been sent as a spy. She strongly disapproves of Bern’s affair with Seamus, and especially of her sharing food, knowledge and hunting skills with him. When he actually disappears two months later, Bernadette nevertheless maintains that knowledge about survival should be shared: “it wasn’t ours to keep” (37). The incident becomes a parable of social ethics; illuminating that every change in even just the smallest community brings about new ethical challenges of interaction. It turns out that Elena had been right about her suspicion: Seamus reappears in May trying to win back the women’s trust with a fresh fish (a “Trojan fish,” as Bern suspects [42]), but he now reveals his actual motivation: He had indeed been sent with a mission “to take whatever you had” (42) and to enable Laird to dispossess or even kill the elders. However, having experienced Bernadette’s generosity, he changed his mind, and

> when I got back, things were worse. Laird wanted to raid you, take everything. I told him I wouldn't bring him here. He couldn't have done it anyway. He was sick, lots of the villagers were, people were dying. The ones who were healthy, we started to trap, to work for the village. Making soup, working in teams. People started getting better. Some didn't. Laird didn't. We nursed him, to death. When he died, we started to talk about new leaders. And—elder—leaders. (42)

When it turns out that he brought Elena's daughter and grandson along to "talk about possible futures" (43), the women begin to cook for the visitors, about to ask everyone into their log cabin.

The social ecology of greed and betrayal, the play underlines, has to remain unsuccessful. In the end, even when only one person upholds the ethics of sharing, trusting, and generosity, it is these ethics that prevail and ensure survival. The ecological collapse that gives the play its title is presented as a logical result of human misconduct, but in this light, the end of the previous world might also be read to mark a new beginning. Even early in the play, Seamus considers the apocalypse an opportunity for personal improvement: "[i]t's like, the—the unplugging—kind of wiped me clean, and I can start learning all over again. But important things this time, not just video games and twitter, but things like what time of day it is from the sun, and what we can eat from the land" (28).

He does learn these things in the course of the play, thanks to Bernadette's kindness, but again, the theatrical engagement with nature is not as simple as it first may seem. The plot is not driven by nostalgia or reductive, back-to-nature naiveness. And just as nature is not idealized in the least, neither is the social ecology of their traditional community: when the two women talk about their exile, Bern brings up that banishment used to be a survival strategy for their people for centuries: "In tough times, —we—used to leave people behind" (10). Elena refuses to approve of this; pointing to the fact that the tribe's nomadic times were long gone, and arguing that their community walled itself in, instead of moving on: "It is still there, building walls and gathering up guns to point at whoever they don't want inside the walls" (10). In the exemplary set-up of a global apocalypse, the exclusion of the two elders is thus not a culturally specific factor of sustainability, but a strategy of empowerment for a privileged few, and thus ethically questionable by universal standards.

Once again, Billy Stratton's model of the "heteroholistic" order—in which Native knowledge (the knowledge about how to hunt rabbits or moose, as well as the knowledge of their language) "emerges from specific geographic locations that are universally valid" (Stratton 149)—provides a framework for the cultural ecology of *The Unplugging.* This framework also extends to the temporal level: just as Marie Clements's *The*

Edward Curtis Project smoothly transcends past and present to position a message about survival and the future, Yvette Nolan's play departs from linear historiography in a temporally specific yet simultaneously timeless setting. This is most obviously reflected in the structure, which roughly covers a time span from December to May: four of the thirteen scenes feature neither plot nor dialogue but simply indicate that "Time passes, moving from Little Spirit Moon, December, to Spirit Moon, January [in scene 3]. The passage of time can be indicated by movement and light" (11). This stage direction is repeated in scenes five, eight, and twelve; with different moons and their approximations into Western months: moving from "Spirit Moon, January, to Bear Moon, February" (17), from "Bear Moon, February, to Snow Crust Moon, March" (31), and from "Pink Moon, April, to Sucker Moon, May" (38), in which the play ends openly. Through these translated (or original, in the case of *mukwa geezis*, or Bear Moon) versions of Anishinaabe months, the play presents different modes of measuring time side by side; enriching the cultural ecology of the stage with an alternative model to simple, linear sequence. All time levels—of past, present, and future—are important for the construction of individual and communal identities: before the characters are reunited with their community, Seamus asks them to "[w]ait, just a moment. I want to remember—everything" (44); and the play ends on the stage direction "They wait" (44). Literally stopping between the memory of the past—which has enabled them to survive by merging traditional Indigenous knowledge with modern Western knowledge (whether it is from contemporary medicine or Robert Frost's poetry)—and the "possible futures" (43) they are about to design, the play ends optimistically on a moment of radical presence. The terminal threat of the apocalyptic event notwithstanding, *The Unplugging* promotes a concept of temporality that Elena phrased earlier, and which strongly impacts characters' identities, their choices and responsibilities: "We are becoming, Bernadette. Becoming who we are" (33).

VI. Conclusion

Developing cultural strategies to cope with environmental change is one of the primary challenges of the twenty-first century. For the communication of efficient policies to protect and share resources globally, across all kinds of separating boundaries, it is crucial to establish interculturally sensitive understandings of specific environments and their impact on human identity. In order to join human beings of diverse cultures in the shared responsibility of sustaining our planetary eco-system, a respectful dialogue between different cultures and systems of knowledge is therefore indispensable. Theater is but one medium to begin this dialogue, but as a protected, accessible, and spatially contained site of experimenting with different ecological understandings, layouts, and changes, it can serve as a laboratory for developing the cultural ecologies of the future.

Whereas much environmental criticism has been "heavily thematic" (Morton *Ecology* 2), the model that I propose in this essay engages with the full semiotic possibilities of the term "ecology;" using it not only as a descriptor for things 'out there,' in a distant, separate, largely projected 'nature,' but as a structural and methodological matrix on which future ecocriticism could be based. Cultural ecology, as initiated by Bateson, Paulson, and Zapf, understands culture itself as a system of interrelated elements, much in the sense that Timothy Morton calls for in *The Ecological Thought*:

> A truly ecological reading practice would think the environment beyond rigid conceptual categories—it would include as much as possible of the radical openness of the ecological thought. Ecocriticism has overlooked the way in which all art—not just explicitly ecological art—hardwires the environment into its *form*. [...] Ecology permeates all forms. (Morton *Ecological* 11)

Read in this way, cultural ecology can be understood as a practice of reading texts and other works of art—performative, visual, or narrative—not only for the strategies in which they construct environments, but as cultural environments in their own right, in which signifiers and codes are interconnected in complex systems of meaning-making. By combining this theoretical concept with the acknowledgement of heteroholism, as a sensitive and respectful model of approaching Indigenous North American texts, performances, and other forms of cultural expression, I hope to contribute to the fruitful crossings of ecocritical engagement, dramaturgy, and Indigenous Studies that are at the heart of this collection of essays.

As the examples by Marie Clements and Yvette Nolan demonstrate, heteroholistic cultural ecology may be used as a viable, inclusive methodology to explore human interactions with their physical, social, and cultural environments—especially in Indigenous or intercultural contexts. If we recognize culture itself as an ecosystem, the stage may serve as a paradigmatic site of identity construction within specific spatio-temporal and spiritual landscapes: nowhere else in art do we find a more densely woven node between local space, subjectivity, presence, and performative agency. First Nations plays are thus particularly telling seismographic registers of larger cultural ecologies, as they negotiate alternative knowledges within changing, vibrantly intercultural environments.

In line with the sideways approach of Eija-Liisa Ahtila's spruce tree at the Kiasma museum, Marie Clements's *The Edward Curtis Project* and Yvette Nolan's *The Unplugging* offer alternative views and fresh perspectives on the relationships between human beings and their natural, social, and cultural environments. These environments are staged as adaptable and dynamic; following Paula Gunn Allen's famous phrase that "[t]he traditional tribal concept of time is of timelessness, as the concept of space is of multidimensionality. In the ceremonial world the tribes inhabit, time and space are mythic" (69). Within these permeable spaces and temporal environments, human identities—both Native and non-Native—are constructed as fluid and flexible, far from any essentialism, and with the full responsibility of acknowledging their ecological embeddings. On an obvious level, the plays leave no doubt about the colonial fall-out, which continues to affect First Nations people all across Canada today: in their natural, political, economic, and cultural ecologies, from the destruction of tribal lands by the tar sands mining or logging industries to the consequences of poverty, and from the pollution or extermination of natural resources to the long-term effects of cultural genocide. At the same time, just as Ahtila's tree in *Horizontal* illustrates the gaps and odd angles, foregrounding the non-mimetic aspects of the artist's solution, the two plays by Clements and Nolan also deconstruct any reductive or naïve approach to nature. The harsh, icy landscapes they present are deathly; there is no space here for Romantic projections of Western desire. In the specific outlines of space, time, and identity, the assumption of a "natural" harmony is not the point—neither in a biological, nor in a cultural sense. When Angeline retells her story communally to acknowledge the systemic violence behind the frozen children, and when Seamus learns that sharing is a much better survival strategy than individual greed, both plays underline an ethics of interconnectedness and mutual respect—which provides the contours for precisely the cultural ecologies that we need in order to sustain the natural ecologies on which our planet continues to depend.

Works Cited

Adamson, Joni. *American Indian Literature, Environmental Justice, and Ecocriticism: The Middle Place.* Tucson: U of Arizona P, 2001.

—, and Scott Slovic, eds. *Ethnicity and Ecocriticism.* Special Issue of *MELUS* 34.2 (Summer 2009).

Ahtila, Eija-Liisa. *Horizontal.* 2011. Kiasma Museum of Contemporary Art, Helsinki, Finland.

—. "Horizontal, 2011." *Kiasma Exhibitions: Eija-Liisa Ahtila.* Kiasma Museum of Contemporary Art, 18 April 2013. Web. 12 July 2013.

—. "Parallel Worlds: Exhibition Publication." *Finnish National Gallery Webstore,* 2013. Web. 12 July 2013.

Allen, Paula Gunn. "The Ceremonial Motion of Indian Time: Long Ago, So Far." Ed. Hanay Geiogamah and Jaye T. Darby. *American Indian Theater in*

Performance: A Reader. Los Angeles: UCLA American Indian Studies Center, 2000. 69-75.

Bateson, Gregory. *Steps to an Ecology of Mind.* Chicago: U of Chicago P, 1972.

Bauerkemper, Joseph. "Narrating Nationhood: Indian Time and Ideologies of Progress." *Studies in American Indian Literatures* 19.4 (Winter 2007): 27-53.

Bennett, Susan. "3-D A/B." Ed. Sherrill Grace and Jerry Wasserman. *Theatre and AutoBiography: Writing and Performing Lives in Theory and Practice.* Vancouver: Talonbooks, 2006. 33-48.

Buell, Lawrence. *The Environmental Imagination: Thoreau, Nature Writing, and the Formation of American Culture.* Cambridge: Harvard UP, 1995.

—. *The Future of Environmental Criticism: Environmental Crisis and Literary Imagination.* Malden: Blackwell, 2005.

Clements, Marie, and Rita Leistner. *The Edward Curtis Project.* Vancouver: Talonbooks, 2010.

Couture, Selena. "Frames of Mind: *Beyond Eden* and *The Edward Curtis Project.*" *Alt.theatre: Cultural Diversity and the Stage* 8.2 (December 2010): 10-17.

Curren, Randall. "Defining Sustainability Ethics." Ed. Michael Boylan. *Environmental Ethics.* 2nd ed. Malden, MA: Wiley Blackwell, 2013. 331-45.

Däwes, Birgit, ed. *Indigenous North American Drama: A Multivocal History.* Albany: State U of New York P, 2013.

Darby, Jaye T. "Into the Sacred Circle: Homecomings in Native Women's Contemporary Performance." Ed. Hanay Geiogamah and Jaye T. Darby. *American Indian Performance Arts: Critical Directions.* Los Angeles: UCLA American Indian Studies Center, 2010. 55-77.

Dreese, Donelle N. *Ecocriticism: Creating Self and Place in Environmental and American Indian Literatures.* New York: Peter Lang, 2002.

Estes, James A., David O. Duggins, and Galen B. Rathbun. "The Ecology of Extinctions in Kelp Forest Communities." *Conservation Biology* 3.3 (September 1989): 252-64.

Frake, Charles O. "Cultural Ecology and Ethnography." *American Anthropologist* 64.1 (1962): 53-59.

Geiogamah, Hanay. *New Native American Drama: Three Plays.* Norman: U of Oklahoma P, 1980.

Glotfelty, Cheryll. "Introduction: Literary Studies in an Age of Environmental Crisis." Ed. Cheryll Glotfelty and Harold Fromm. *The Ecocriticism Reader: Landmarks in Literary Ecology.* Atlanta: U of Georgia P, 1996. xv-xxxvii.

Grinde, Donald A., and Bruce E. Johansen. *Ecocide of Native America: Environmental Destruction of Indian Lands and Peoples.* Santa Fe, NM: Clear Light, 1995.

Heise, Ursula K. "The Hitchhiker's Guide to Ecocriticism." *PMLA* 121.2 (2006): 503-16.

—. *Sense of Place and Sense of Planet: The Environmental Imagination of the Global.* Oxford: Oxford UP, 2008.

Holm, Tom, J. Diane Pearson, and Ben Chavis. "Peoplehood: A Model for the Extension of Sovereignty in American Indian Studies." *Wicazo Sa Review* 18.1 (2003): 7-24.

Huggan, Graham, and Helen Tiffin. *Postcolonial Ecocriticism: Literature, Animals, Environment.* London: Routledge, 2010.

Kidwell, Clara Sue. "Native American Systems of Knowledge." *A Companion to American Indian History.* Ed. Philip J. Deloria and Neal Salisbury. London: Blackwell, 2002. 87-102.

Krech, Shepard E. *The Ecological Indian: Myth and History.* New York: Norton, 1999.

Leistner, Rita. "Artist Statement." Marie Clements and Rita Leistner. *The Edward Curtis Project.* Vancouver: Talonbooks, 2010. 71-73.

Matthiessen, Peter. *In the Spirit of Crazy Horse.* New York: Viking, 1983.

May, Theresa J. "Beyond the Nature/Culture Divide: Challenges from Ecocriticism and Evolutionary Biology for Theater Historiography." Ed. Henry Bial and Scott Magelssen. *Theater Historiography: Critical* Interventions. Ann Arbor: U of Michigan P, 2010. 148-61.

—. "Greening the Theater: Taking Ecocriticism from Page to Stage." *Interdisciplinary Literary Studies: A Journal of Criticism and Theory* 7.1 (Fall 2005): 84-103.

Mojica, Monique. "Chocolate Woman Dreams the Milky Way." Ed. Birgit Däwes. *Indigenous North American Drama: A Multivocal History.* Albany: State U of New York P, 2013. 123-39.

Morton, Timothy. *Ecology Without Nature: Rethinking Environmental Aesthetics.* Cambridge: Harvard UP, 2007.

—. *The Ecological Thought.* Cambridge, MA: Harvard UP, 2010.

Nelson, Robert. *Place and Vision: The Function of Landscape in Native American Fiction.* New York: Peter Lang, 1993.

Nolan, Yvette. *The Unplugging.* Premiered October 17, 2012 at the Arts Club Theatre, Vancouver. Unpublished Manuscript. Courtesy of the Playwright.

Paulson, William. "Literature, Knowledge, and Cultural Ecology." *SubStance: A Review of Theory and Literary Criticism* 22 (1993): 27-37.

Ruppert, James. "Review of *Two Old Women*, by Velma Wallis." *Ethnohistory* 42.4 (Fall 1995): 670-71.

Rutherford, Stephanie. *Governing the Wild: Ecotours of Power.* Minneapolis: U of Minnesota P, 2011.

Schweninger, Lee. *Listening to the Land: Native American Literary Responses to the Landscape.* Athens: U of Georgia P, 2008.

Shorty, Robert. *Na Haaz Zan.* 1972. Premiered October 25, 1973. La Mama Experimental Theatre Club, New York. Unpublished Manuscript. New York Public Library, Lincoln Center for the Performing Arts.

Soper, Kate. *What is Nature? Culture, Politics and the Non-human.* London: Blackwell, 1995.

Steward, Julian Haynes. *Theory of Culture Change: The Methodology of Multilinear Evolution.* U of Illinois P, Urbana 1955.

—. "The Concept and Method of Cultural Ecology." Ed. Nora Haenn and Richard Wilk. *The Environment in Anthropology: A Reader in Ecology, Culture, and Sustainable Living.* New York: New York UP, 2006. 5-9.

Straight, Nathan. *Autobiography, Ecology, and the Well-Placed Self: The Growth of Natural Biography in Contemporary American Life Writing.* New York: Peter Lang, 2011.

Stratton, Billy J. "Towards a Heteroholistic Approach to Native American Literature." *Weber: The Contemporary West* 29.2 (Spring/Summer 2013): 148-50.

Thornton, Russell. *American Indian Holocaust and Survival.* Norman: U of Oklahoma P, 1987.

Wallis, Velma. *Two Old Women: An Alaska Legend of Betrayal, Courage and Survival.* New York: Harper Perennial, 1993.

Walters, Anna Lee. *Talking Indian: Reflections on Survival and Writing.* Ithaca, NY: Firebrand Books, 1992.

Weaver, Jace. *That the People Might Live: Native American Literatures and Native American Community.* New York: Oxford UP, 1997.

Woody, Elizabeth. "Voice of the Land." Ed. Simon J. Ortiz. *Speaking for the Generations: Native Writers on Writing.* Tucson: U of Arizona P, 1998. 148-73.

Zamir, Shamoon. "Native Agency and the Making of *The North American Indian:* Alexander B. Upshaw and Edward S. Curtis." *American Indian Quarterly* 31.4 (Fall 2007): 613-53.

Zapf, Hubert. "The State of Ecocriticism and the Function of Literature as Cultural Ecology." Ed. Catrin Gersdorf and Sylvia Mayer. *Nature in Literary and Cultural Studies: Transatlantic Conversations on Ecocriticism.* Amsterdam: Rodopi, 2006. 49-70.

Mounds, Earthworks, Side Show Freaks and Circus Injuns

Ric KNOWLES[1]

University of Guelph

In Summer of 2012, I was invited to take part in a panel at the Association for Theatre in Higher Education in which we were asked to discuss what we viewed as "foundational documents" for theatre in America. The conference took place in Washington, DC, a city famous for its monumental architecture, most notably the neoclassical grandeur of the Capitol Building, the White House, the Jefferson Memorial, and the US Supreme Court—all, in their domes, classical columns, and domination of the landscape, evoking the European "enlightenment,"[2] a moment in history that was not kind to the world's Indigenous population.

Just down the Potomac from the capital, on both sides of the river, near Mount Vernon (the estate of George Washington) but beyond the awareness of most politicos and tourists alike, there are different and much older kinds of monumental architecture, structures with a far different relationship to the land. These, and the thousands of other ancient Indigenous mounds and earthworks that populate Turtle Island (North

[1] This essay could not have been written without the collaboration of my colleagues on the "Indigenous Knowledge, Contemporary Performance" research project, funded in part by the Research/Creation program of the Social Sciences and Humanities Research Council of Canada, including: Omushkego Cree playwright Candace Brunette, Anishinaabe scholar and performer Jill Carter, anthropologist Brenda Farnell, Choctaw novelist, poet, playwright and essayist LeAnne Howe, Plains Cree director and actor Michael Greyeyes, Omushkego Cree designer Erika Iserhoff, and Kuna/Rappahannock playwright, performer, essayist and editor Monique Mojica. My account of the mounds also draws heavily throughout on the work of Chadwick Allen.

[2] Although the neoclassical style dominates US federal buildings on and around Capitol Hill and the Mall, there are many distinctive architectural styles on view in the US Capital, including the distinctively different National Museum of the American Indian, designed by Douglas Cardinal, a Blackfoot architect from what is now Canada, which sits uncomfortably as a "national" museum among others on the mall, but carves out its curvilinear difference amidst 4.25 acres of landscape recreating various ecosystems in the Americas.

America), were what I chose to discuss as quite literally foundational for American theatre and performance history: performances were held on them, they themselves constituted performances, and they remain the dramaturgical basis for an Indigenous performance-in-progress that this essay will describe, modeling a different way of thinking about theatrical process, product, and relationship to the ecosphere than regularly obtains in western dramaturgies.

A lengthy discussion followed the Washington panel, with many questions and comments from the floor, but none addressed my presentation, none acknowledged the mounds. This is not entirely surprising. Few in the western academy, myself included, know what *kinds* of questions to ask of these monumental but somehow invisible earthworks, which sit on the landscape as they did in the air after my presentation, apparently inscrutable. This essay might serve, then, as a kind of progress report—or as a report on a process—that chronicles my experience as a non-Native dramaturge working with Indigenous artists who are attempting to listen to and honour the wisdom of the mounds themselves.

My choice of the mounds began with a premise: that foundational to performance on Turtle Island are the foundational performances of the first *peoples* of Turtle Island, performances that are too often forgotten or relegated to "pre-history," researched and catalogued by archeologists rather than theatre historians, and by scholars whose relationship to the land is the secondary one of the settler/invader. Even much contemporary Indigenous theatre fails to take into account the foundational performance forms and cultural texts Native to this place, this land—in contemporary parlance, this ecosphere.

I offered my chosen foundational documents as a corrective to this: the earthworks, platform, conical, pyramid, ridge-top, geometric, burial and effigy mounds of the Indigenous peoples of Turtle Island that date back millennia, and reach across much of the continent. The building of these mounds was performative, and they served as the sites of ceremony and performance for the peoples who first built them in the Archaic era (8000-2000 BCE) and the Woodland period (1000 BCE-1000 CE), in the Coles Creek (700-1200 CE), Mississippian (900-1450 CE), Fort Ancient (1000-1650 CE), and Plaquemine (1200-1750 CE) cultures; they are experiencing a revival today, being restored and even newly built, and they are serving as structural models for artistic, literary, theatrical, and dramaturgical practice among many contemporary Indigenous nations.

The mounds are remarkable. Although they have suffered varying degrees of desecration or "preservation," they not only survive, but are alive and talking to Native peoples throughout Turtle Island. There are mounds from Brazil to northern Manitoba in Canada, but the major extant mound sites stretch north from the Lower Mississippi Valley, across the Ohio Valley, and into Ontario. At their peak, tens of thousands of mounds dotted Indigenous North America, beginning as early as 4000 BCE, and they functioned not to dominate the land, but as a means of living in harmony with it. The very name "Turtle Island" connotes a vast effigy mound rising out of the oceans.

And the mounds have a very precise relationship, not only to the surface of the natural world, but also to dimensions above and below the earth plane. As Native American scholar Chadwick Allen says, "[e]arthworks parallel natural embankments, follow the waterways of rivers and creeks, mirror the regular seasons of the sun and moon, the patterned coordinates of stars in transit. They occupy symbolic positions, too, aligning within North American systems of representation, within complexes of ceremony and ritual, with economies of power and exchange" (Allen "Serpentine" 815-16).[3] At the time of their construction, many of the mounds were only legible from a then impossible height, either in their totality as effigies, usually of birds, animals, or reptiles, or in their relations to one another: geometric mounds sixty miles apart near Newark, Ohio, for example, were designed to "nest" inside one another, but were linked by straight two-hundred-foot wide "star roads" that were aligned at certain times of the year with the Milky Way. They were, it seems, not intended for human viewing (Allen "Serpentine" 811).

The building of the mounds, formed of interlaced layers of different types, colours, and textures of soil imported from different locations, required a precise knowledge of mathematics, physics, engineering, astronomy, and architecture, all understood as "natural" sciences. And as non-Native archeo-astronomer Anthony Avilo suggests, these mounds constitute "a marvellous way of connecting the ecosphere, the ecological environment, with the heavens. And maybe we modern astronomers tend to cut off the heavens from the rest of the world—the mountains, the water, the plants, the animals—to which I think most ancient cultures would have forged a connection" (qtd in *Earth Works*). Indeed, as Tewa (Pueblo) ethnobotanist Gregory Cajete argues, "[t]he historic efforts of Native cultures to resonate with the heavens also represent their attempts to live up to an ideal ecological relationship with the Earth" (Cajete 256).

The mounds performed many different functions in the ancient world, marking territorial boundaries and roadways, creating focal points for trade, economic, and artistic exchange, and serving as sites for intellectual and spiritual practice. Some of the platform mounds, such as the massive,

3 Although not an enrolled citizen of the Chickasaw nation, Allen traces Chickasaw ancestry through his maternal line (*Trans-Indigenous* 250 n11). The Chickasaws themselves trace their ancestry to Mound Builders.

13.8 acre, 15' high, terraced Monks Mound at the Cahokia site in Illinois, may have been constructed over centuries and served as foundations for structures, games, and ceremonial practices. Others, such as the majestic Serpent Mound in Ohio, the largest serpent effigy in the world at 1370 feet, include figural representations and multiple astrological alignments, and appear to have been built to maintain cosmic balance among the heavens, earth, and underworld (Allen "Serpentine" 817-19). As Allen argues, the mounds

> represent achievements in science and aesthetics on a monumental scale. They integrate the precise observation of natural phenomena with geometry and other abstract forms of knowledge, as well as with practical skills in mathematics, architectural design, engineering, and construction. Many earthworks were sculptured to mirror perceived patterns in the sky, both in the bodies of individual works and in the arrangements of multiple works into complex sites and cities; moreover, particular works were often aligned with specific celestial events, such as an equinox or solstice sunrise or sunset point on the horizon. (Allen 809)

Many of the once thousands of mounds have been looted and/or destroyed, and many are still hiding in plain sight in rural and urban areas across the continent, often unknown to those who live closest to them. But there are prominent exceptions, mounds that have been largely preserved and subjected to various kinds and degrees of historical, anthropological, astronomical, and archeological interpretation and analysis by Native and non-Native scholars. Among the most prominent are the Cahokia Mounds in Illinois, including the monumental so-called Monk's Mound; various serpent mounds, most prominently the Great Serpent Mound at Peebles, Ohio, but also including the threatened sacred snake mound at Blood Run on the Big Sioux River on the Iowa-South Dakota Border, the serpent mound in Toronto's High Park that has been nearly destroyed by BMX bike riders, and the serpent mound on the shore of Rice Lake, near Peterborough, Ontario; and the astonishing and mysterious constructions at Chaco Canyon, New Mexico, now a national historic park. All of the mounds and earthworks are located near water, all are aligned with significant solar or lunar events, and all are closely oriented to the natural lay of the land.

Chadwick Allen, in an essay about Alison Hedge Coke's book of poems, *Blood Run*, based and structurally modelled on the mounds at the Oneata site on the Big Sioux River, argues that the mounds constitute "a form of Indigenous *writing* that employs 'geometric regularity' and 'geometrical harmony' in order to record natural, human, and cosmic relationships within durable structures" (812). He uses as an example the Great Serpent Mound in Ohio:

> The body of the extant Serpent Mound runs roughly a quarter of a mile in length and is mounded to a height of about three feet. It begins at its southern

> end with its tail in a triple-coiled spiral, undulates along the plateau above the Brush Creek Valley in seven distinct body convolutions, and then straightens out toward its broad, horned head. The Serpent's mouth hinges wide open, poised to swallow a large, oval-shaped disk. ("Serpentine" 817)

Drawing on the work of anthropologist William F. Romain, Native scientist Gregory Cajete, and others, Allen demonstrates that the mound has eight astrological alignments (true astronomical north, the summer solstice sunset point, and six lunar rise and set points on the horizon), and that it embodies the philosophical and spiritual concept of cosmic balance (818). "The Mound Builders," according to Romain, "celebrated world renewal ceremonies at the site, in order to help strengthen the powers of the upperworld in the continuing struggle against the forces of the underworld" (qtd in Allen 818), and according to Cajete this resonates with Indigenous astronomy more broadly as the peoples' attempt "to align themselves and their societies with what they perceived was the cosmic order" (Cajete 217).

For Allen the mounds constitute "Indigenous writing systems" (829)—the expressive literature of the Indigenous peoples who created them. In a contemporary world in which Native cultures are most often considered to be preliterate, in spite of winter counts, Mayan Codices, inscribed Passamaquoddy birch bark scrolls, petroglyphs, pictographs, wampum, and other forms of writing, this is important.

But Choctaw writer and scholar LeAnne Howe says that the mounds also constitute *performances*, and they are performances with a very clear relationship to the rhythms of the natural world from which they emerge. She provides the example of the Great Bird Mound at the extensive Poverty Point site near West Carroll Parish, Louisiana, an ancient long-distance centre for trans-Indigenous trade and commerce. At seven stories high the Bird Mound is the Western hemisphere's largest Archaic period earthwork, and second in size only to Mississippian-era Monk's Mound, yet this spectacular bird was created in only three short months. The builders moved some 238,000 cubic meters of soil in 50-pound baskets in just 90 days, Howe argues, as a performance of the creation story of the red-tailed hawk, whose *own* gestation period from conception to first flight is also 90 days. Both the building of the mound and the conception of the hawks begin at the vernal equinox, and the completion of the mound coincided, three months later, at summer solstice, with the fledgling hawk's first flight. Both solstices, moreover, would subsequently have been celebrated at the mound site.

> I think the builders of the site were performing the creation story of a literal bird. If the Bird Mound is itself a kind of theatrical performance, one in which the performers tell the story by collectively sculpting the earth, it calls

> into question the very nature of mound building as, "writing on the land." Bird Mound, as a creative performance, affirms collaboration as an ancient indigenous epistemology, perhaps considered the highest art form expressed by indigenous builders, artists, and players that carried the soil in baskets. (Howe and Wilson 5)

But these mounds did not only constitute *ancient* performances; in fact, as I write, mounds are being restored and a new one is being constructed by the Oklahoma tribes at the American Indian Cultural Centre in Oklahoma city in a project designed explicitly to participate in a centuries-long tradition by "expanding the vocabulary of earthworks as an iconic landscape" (Wasserman 2, qtd in Howe and Wilson 25ff). I would argue, moreover, that the extant mounds continue to perform their peoples across time, to the point that Howe can wonder, "are we, as tribal/indigenous peoples, the *embodiment* of mounds sites?" (Howe, "Comes Now" 6, emphasis added). This mound performance manifests itself in poetry, as in Alison Hedge Coke's *Blood Run*; in the visual arts, as in the paintings of America Meredith (Cherokee) and the digital/mixed media work of Alyssa Hinton (Tuscarora); and finally in the theatre, as in *Side Show Freaks and Circus Injuns*, a research/creation project involving LeAnne Howe and Kuna/Rappahannock Monique Mojica as writer/performers, Plains Cree Michael Greyeyes as director, designers from various Indigenous nations in what are now the U.S. and Canada, and myself as dramaturge.

Side Show Freaks listens to and attempts to embody the mounds, and uses their architectures and construction processes as its foundation and dramaturgical model. Indigenous artists constructing a dramaturgy grounded in the architecture of the mounds reconnect and return Indigenous memories, in the form of performances—and all cultural memory is performative—to the present world. It addresses in its content a history (based on the family backgrounds of Mojica and Howe) of the exploitation and display of Indigenous peoples, who have, like the mounds, been at once invisible and hypervisible, overlooked and yet also subjected to the scientific, scopophilic, or pornographic gaze. But the show is employing the environmental and astronomical optics of the mounds to construct, as its foundation, a contemporary Indigenous dramaturgy that at once returns that gaze and reassesses the peoples' relationship to the ecosphere and the universe.

Side Show Freaks and Circus Injuns is in the early stages of its development.[4] Envisioning and early writing for the project began in

[4] The process is long and arduous—it takes years and many workshops to bring a show to completion, as it took years to build many of the mounds—and the team interrogates each step closely to ensure that we remain true to the dramaturgical principles that we have distilled from the mounds and earthworks that are our models, not slip into the

2005, with writing Mojica did as "out-takes" on her project, *The Only Good Indian*, with Turtle Gals Performance Ensemble in Toronto. But research began in earnest in 2011 as a result of an International Development Grant from the Ontario Arts Council, and a Research Creation Grant from the Social Sciences and Humanities Research Council of Canada. From June to October of that year, and on specific occasions since then, members of the research team have researched and visited most of the major mound sites and many of the lesser ones (see Mojica "In Plain Sight"), engaging in "embodied research," a process of making kinesthetic connections with the mounds and the sacred sites that they inhabit. They have brought that research in their bodies to a creation workshop in June 2012 where we engaged in a process of "deep improvisation"—embodied writing—followed by analysis, rewriting, and refining, crafting the material into a performance score. As Mojica describes it,

> I practice an Indigenous artistic research methodology that speaks to the embodiment of place. The land *is* our archive and our embodied relationship to the land defines Indigenous identities, history, science, cosmology, literature—and performance. [...] *Side Show Freaks and Circus Injuns* [...] solidifies the location of Indigenous performance principles at the centre of my artistic practice. It regenerates my creative source by placing me on the life-giving land in an embodied research process that requires me to walk on, touch, feel, smell, and absorb the stories, forms, and structures of effigy mounds and earthworks to connect to the ancestors who built them and to the peoples who still inhabit the region. Simultaneous to this sacred work, I am challenged to "talk back" to colonial erasure, to peel away that veil, to refocus my lens. Once we were not victims. (Mojica "Artist Statement")

The first week of the June 2012 workshop was a process of distilling our research, embodied and otherwise, into basic principles, knowing that we had to invent a creation process analogous to that of the Mound Builders themselves. In creating this play, we had decided, we were, in effect, building a mound. This, we determined, involved:

1. layering and texturing of materials and "ribboning" ourselves as a collaborative creation group engaged in process *as* performance;
2. connecting and aligning ourselves and our work with the land, the underworld, and the starworld, mapping our process according to the Mississippian-era cosmological model, "a three-level

---

taken-for-granteds of dominant theatrical practice in the west. The first public presentation of *Side Show Freaks and Circus Injuns* of any kind was at the Canadian Federation for Advanced Research "Belonging Differently" meetings in Banff, Alberta, in August 2012, but we will continue the mound-building process until the premieres in the early months of 2015.

configuration composed of the Above World or Overworld (i.e. the sky), the Middle World, and the Beneath World or Underworld," with a central axis connecting all these levels in the form of "a center pole or sacred tree" (Reilly 127);

3. thinking durationally, thinking in eons, in "deep time," and considering the relationship between our development time, the duration of the performance, and the duration of its afterlife, its remains;
4. building a *ceremony* of moundings[5]; and
5. maintaining a kind of processual openness that would allow the mounds, in a very real sense, to perform us.

Once in studio we began by doing creation exercises that explored duration ("deep time") and ribboning: ribboning bodies, sound, and imagery even as the Mound Builders ribboned different colours, textures, and densities of earth, often brought from great distances. Director Michael Greyeyes (Plains Cree) introduced an exercise in duration, beginning with familiar actorly exercises of breathing and connecting to the floor. As he talked Mojica and Howe through the exercise—"Your duration is of generations. Your duration is measured in eons"—he passed me a note: "I'm trying to connect them to deep time." "Even if your movement is sudden, like an earthquake" he intoned, "it was one million years of shifting that produced it. Your breath is the wind that shaped the landscape. You are earth and ground and rocks"—another note, "they are mounds,"—"but that doesn't mean that you move glacially. You always move gently, and it is formed by breath, the way the wind changes the dunes from night to night. The landscape rolls; you may roll. You are a continent, and you move because your breath moves." This last became a repeated refrain. "You move because your breath moves. As you move, parts of you move internally. Sediment shifts, even if there may be no external movement visible. Your breath is the duration of an epoque. In one breath monkeys became bipedal. Then you gently shift back into where you entered this […]. You breathe, you shift. Your breath will bring you home, like a glacier that melts, like a rock that weathers, you find a beginning place. Your movement is the language of duration."

As the actors emerged from the exercise, we discussed its applications and the questions that it raised: how is duration expressed in movement? How does it affect our language? What words speak duration? What words take an eon to say? What is the duration of this space, this place? What

[5] We are influenced in our thinking by Opaskwayak Cree scholar Sean Wilson's book, *Research is Ceremony: Indigenous Research Methods*, and we consider not only our research but also our creation process to constitute ceremony.

is the language of duration? We talked about the rhythms of the show as intercutting two different time zones, two different durationalities, those of the mounds (the sacred) and those of the side shows (the profane), and this led us to an awareness of the side show chronotope as one that, unlike that of the mounds, is not alive, does not breathe, and has no organic relationship to place. It is the chronotope of pornography.

A second time-travel exercise ensued, one designed to make the bodies of the performers experience and endure "mound time," getting to know, too, what we came to call "the stealth nature of mounds"—their capacity, over eons, subtly to commune with contemporary consciousness. Greyeyes asked the performers to lie on the ground next to a chair, and to take a full twenty minutes to move continuously and purposefully—"because the mounds are purposeful"—into a seating position on the chair. The exercise was extremely difficult, "making," as Greyeyes said, "the body endure the interminable time function: what happens to the body in mound time as it moves in space." Mojica imagined the mound itself having an itch and needing to scratch for six hundred years, waiting for someone—us—to visit and bring relief.

When we moved into improvisation/creation mode we began by reading a brief excerpt from an account Mojica had published about her research travels to mound sites as a way of reconnecting with our research:

> *Across the thirteen moons on Turtle's back, across multiples of centuries the Mother Mounds are calling to their children: "Come home. ..." They call, sing, coo, echo, hum. They infiltrate our sleeping and our waking dreams with whispers of thoughts they make as if our very own: "Come home ... it's time. Come home to rest your hearts in the layered folds of Mother's skirts." They send out a call.*
>
> *We who think these thoughts as if they were our very own search for the source of the song, the Mother voice, and then gently lay down each boney vertebra of our spines one by one on the ancient, textured soil. Then, vibrating, we align our frequencies to the constellations and join a dialogue begun long ago, mirrored from the Upper World onto nodes and nerve centres on Turtle's back, creating the World Between through earthworks imagined, crafted, and mounded, basket by fifty-pound basket, by minds, hands, thighs, and muscled backs, through Earth itself.*
>
> *We are in conversation. This is real.* (Mojica "In Plain Sight" 218, emphasis in original)

The improvised monologue that emerged from a process that had begun with Mojica lying on her belly on the mounds themselves—communing with them, embodying them—and proceeded through Greyeyes' exercises in durational time, was highly charged. Mojica's improvisation moved her from the Great Serpent Mound to the magnificent effigy mounds at National Monument, MacGregor, Iowa, overlooking the Mississippi River (Little Bear Mound, Great Bear Mound, Fire Point, and Eagle Rock), and from outside the mounds, visiting, to the inside, eventually embodying the mounds themselves, which seemed to speak *her*, rather than the other way around. The rough improvisation is excerpted here:

> We weep, belly down. Sobs. Sobs, heart on the serpent, belly on the snake, we feel it, undulate, beneath our bellies, bringing us its song. Thank you. Thank you, for leaving this for us to see, when all is almost gone, your bones pulverized to dust. […] We reconstitute our broken bodies from your bones… Buzzard watches. Shadow bird. Song begins. Serpent writhes, long ago meteor charged, long ago meteor radiating through our spines: three coils of its tail, seven undulations and release. Then circle. The Great Serpent reaching towards water. Buzzard watches. We weep, we weep, belly down on serpent's body, riding, undulating beneath our bodies. Heart to serpent. Reaching towards water, we weep. […] Reconstitute our bodies from the pulverized bones of our ancestors who left this for us to see when all was almost gone, when almost nothing was left… three coils of its tail, seven undulations… radiating out, never stop never stop never stop, catch: we weep to see this place, three coils of its tail, seven undulations of its body. […] Thank you for leaving this place for us to see when all is almost gone, when there is almost nothing left. […] The song begins, and we who think these thoughts as if they were our very own, answer the call.
>
> It's time. It's time, come home and rest yourself in the folds of mother's skirts. There's something I must tell you, there's something you need to know. Once there was conversation, come closer, come closer, I almost have you. Here. Here, here, my most prized possession, needles made from bird bone tucked under my arm, come, come, come, reaching towards water, looking out over power my most prized possession, my bundle of needles made from bird bone. [*pause*] […]
>
> Mounds of light and shadow. Breathing, breathing, breathing. Built basket by 50 pound basket. Breathing.
>
> AAARGH. They're looking at me. Looking at MEEEE. No flesh to protect from their gaze on my bones. They're looking at me. Blind eyes. Exposed, no flesh, digging, digging. […] Remove the blanket of earth. Looking at me. […] Breathe. I remember. Breathe. The lungs of the earth. Brother, sister tree: Shade. Protects. Stands by. Sentinel. Water: Touches shore, retreats. […] We are still here. We are still here. Endure, remain, endure, remain, remains, remains, enduring remains that tie you here, bring you here. An approach, an approach tickles my spine, an approach crawling down my hip bone, a shift that circles the top of my femur. […] I smell tobacco [*sniffs*], someone breathes and brings tobacco, someone (breathe shift glide, tingle): approach, approach, approach, approach. The sweet tobacco. The tips of my finger bones. Glides and soothes my naked skull. "Who are you?" Come so far, so distant, so new? [*exhales*]. You need to know, you need to see. You are part

> of this knowing, this doing, this breath [*inhales*]. This shift, this moment, this come so far from a place of dreaming, from a place of numb, from a place of loss, from a place of gaps, from a place disjointed. You need to know my breath is your breath. (Mojica, from "Outside/Inside")

If the creation process for *Side Show Freaks* is based on mound building, it stands to reason that the dramaturgical structure of the show that results will be that of a mound. The four structural principles that our collective field research identified as being characteristic of all of the different types of ancient mounds and earthworks are: duration, alignment, convergence, and integration. These principles will constitute the bones of the show-as-mound that we are building.

Dominant western dramaturgical models since Aristotle have privileged three part structures, structures having to do with: beginnings, middles, and ends; rising actions, climaxes and falling actions; theses, antitheses, and syntheses; two character scenes into which a third brings productive complications that advance the action; and so on. These structures support dramatic action that is based on conflict, on the achievement of objectives, and on a myth of linear, progressive time.

Duration, alignment, convergence, and integration constitute a very different deep structure, and serve as dramaturgical principles to support a different dramatic action, one that relies less on conflict and domination, one that embodies the patience of the mounds, and one that reflects a profound effort to operate at a deep level in a negotiated harmony with the ecosphere, with the universe. The mounds, as Mojica argues, "were built with deliberate creative intent, requiring precise knowledge of physics, engineering, astronomy, and architecture. They have been lovingly mounded to carve out ceremonial spaces conscious of their theatrical impact in order to prepare the human body to be aligned in body, mind, spirit, and heart with the cosmos" ("In Plain Sight" 221). We build with the same intent.

Works Cited

Allen, Chadwick. "Serpentine Figures, Sinuous Relations: Thematic Geography in Allison Hedge Coke's *Blood Run*." *American Literature* 82.4 (2010): 807-35.

—. *Trans-Indigenous: Methodologies for Global Native Literary Studies*. Minneapolis: UP of Minnesota, 2012.

Cajete, Gregory. *Native Science: Natural Laws of Interdependence*. Santa Fe: Clear Light, 2000.

Earth Works: *Virtual Explorations of the Ancient Ohio Valley*. Cincinnati, Ohio: U of Cincinnati CERAC. Centre for the Electronic Reconstruction of Historical and Archological Sites, 2007.

Howe, LeAnne. "Comes Now Ballgame's Tribalography: Embodied Story." *Studies in American Indian Literature* (2014), forthcoming.

Howe, LeAnne, and Jim Wilson. "Life in a 21st Century Mound City." Ed. Robert Warrior. *World of Indigenous North Americas*. Routledge, 2014, forthcoming.

Mojica, Monique. "Artist Statement." Draft application to the Ontario Arts Council Creators Reserve play development fund, 13 January 2013. Unpublished ms draft.

—. "In Plain Sight: Inscripted Earth and Invisible Realities." Ed. Roberta Barker and Kim Solga. *New Canadian Realisms.* Toronto: Playwrights Canada, 2012. 218-42.

—. "Outside/Inside." Monologue improvised at the creation workshop for *Side Show Freaks and Circus Injuns*. Cahoots Theatre Company studio, Toronto, Ontario. June 2012. Unpublished ms.

Reilly, F. Kent, III. "People of Earth, People of Sky." Ed. Robert V. Sharp. *Hero, Hawk, and Open Hand: American Indian Art of the Ancient Midwest and South*. Art Institute of Chicago. New Haven: Yale UP, 2004.

Wasserman, Shoshanna. "21st Century Mound Builders." Document Downloads / Past Press Releases. *The American Indian Cultural Center and Museum*. 5 August 2008. Web. 21 March 2012.

Wilson, Shawn. *Research is Ceremony*: *Indigenous Research Methods*. Halifax: Fernwood, 2008.

"Civilization" and its Transgressions on the Old Shawnee Trail

Lynn Riggs's *Out of Dust*[1]

Jaye T. DARBY

University of California-Los Angeles

> It's a large order to be alive in the world today. It's strenuous and often demoralizing. But the more we seek to know and to comprehend, and to add what we can to make it bearable for ourselves and others to live, the more revelation we stumble on, and the more we possess the power to change that world.
>
> Lynn Riggs (Credo 167)

An activist playwright and director throughout his career spanning the late 1920s to early 1950s, Lynn Riggs, an enrolled member of the Cherokee Nation, brought a unique ethical perspective to the North American stage, including Broadway.[2] In 1941 Riggs articulated a transformative vision for theatre as a moral, social, and ecological force

1 My deep gratitude goes to the Institute for the Study of the Americas and School of Advanced Studies at the University of London, where I was a visiting research fellow in 2008. Their human rights ethos and multidisciplinary focus deepened my research project, "Re/Casting the American West: Native Theatre and Broadway Reconsidered," and nurtured initial development of this chapter. Early versions of these ideas benefited from Association for Theater in Higher Education (ATHE) and Comparative Drama conference sessions and audience comments. For their insights and support, special thanks also go to Birgit Däwes, Hanay Geiogamah, Marc Maufort, Courtney Elkin Mohler, Marjorie Faulstich Orellana, Myrton Running Wolf, Michael Seltzer, Christy Stanlake, Concepción Valadez, and Ken Wade. Additionally, I so appreciate the assistance of David Kennedy, Associate Artistic Director; Annie Keefe, Artistic Advisor; and Patricia Blaufuss, Public Relations Manager, at the Westport Country Playhouse, and Paul Heller and Chilton Ryan, who worked on the original production of *Out of Dust*.

2 To date, Braunlich has provided the most complete biography on Lynn Riggs with her book *Haunted by Home*. Two dissertations by Aughtry and Wilson in 1957 and 1959, respectively, based on Riggs's papers and colleagues' first-hand accounts, are also valuable sources.

in "A Credo for the Tributary Theatre," featured in the 25th anniversary issue of *Theatre Arts*, a message he had delivered earlier at the Globe Theatre in San Diego. As suggested by the opening quotation, Riggs located theatre within broader human rights efforts, emphasizing the "right to dignity" (167). According to Cherokee scholar Daniel Heath Justice, "Riggs sought to give a liberated voice to the marginalized throughout his brief but productive career" (92). Drawing on his inclusive vision of humanity, Cherokee heritage, and deep ties to his homeland, Riggs filled his plays, such as *Big Lake*, *A Lantern to See By*, *Roadside*, *Green Grow the Lilacs*, *The Cherokee Night*, and *Out of Dust*, with complex stories and ethical struggles of diverse people, mostly living in Indian Territory and Oklahoma.

These geopolitical distinctions were highly significant to Riggs and Cherokee history. Indian Territory, now part of eastern Oklahoma, was established after the Indian Removal Act of 1830 and brutal Trail of Tears. Under President Andrew Jackson, these governmental actions forced the Cherokee and other nations off their sacred homelands in the southwest to this area, west of the Mississippi River (cf. Justice 60-88). According to historian Angie Debo in her landmark book *And Still the Waters Run*, even though these treaty-protected lands in Indian Territory belonged to the "Cherokee, Choctaw, Chickasaw, Creek or Muscogee, and Seminole nations" (1), pressure from the U. S. Government and European American settlers increasingly threatened them. Debo documents a series of federal actions throughout the late 1800s and early 1900s that systematically dismantled earlier treaties and sought once again to dispossess the Cherokee Nation and neighboring nations of their communal lands through allotment and to strip their sovereignty rights. As a result, on November 16, 1907, Oklahoma became a state, combining Oklahoma Territory and Indian Territory. Far from a day of celebration, historian Rose Stremlau describes the prevailing Cherokee reaction: "The inherent brutality of this union was painful for many Cherokees, and they could not bear witness" (179). Born in 1899 in "Claremore, Indian Territory" (Aughtry 8), Riggs experienced these tumultuous times firsthand.

Thus, Indian Territory in his plays often serves as a repository of Native spiritual, cultural, geographic, and political meaning. Linking Riggs with other activist Cherokees in the early twentieth century—John Milton Oskison, Will Rogers, and Emmet Starr, Justice stresses all four men's "geographic connectedness to their work, a consciousness and conscience rooted in the land of their birth" (93). He continues: "As citizens of the Cherokee Nation and then inheritors of the statehood legacy of Oklahoma, Riggs, Oskison, Rogers, and Starr reflected on the injustices inflicted by White Americans on Indian communities, and all sought ways of effectively redefining and reimagining Indian lives, histories, and identities away from the self-serving distortions of dominant White political and social systems" (94). Their work and that of many others sustained the Cherokee Nation after statehood.

Over the past decade, initiated by Cherokee scholar Jace Weaver, Lynn Riggs's legacy and work, especially *The Cherokee Night*, have gained significant recognition.[3] A number of analyses focus on the play's connections to Cherokee traditions, lands, and history, especially in terms of the devastating effects of federal policies at the end of the nineteenth and early twentieth centuries[4]. To build on this scholarship and encourage future productions of Riggs's plays, this chapter offers a transformative dramaturgical analysis of *Out of Dust* in 1949, a major but only recently published drama about a European American family on the Shawnee Trail in Indian Territory in the late 1800s, that exposes the ruthless ideology of "civilization" used to justify Westward expansion in the U.S.

Guided by Native critical studies, my approach draws on the works of Pueblo/Métis scholar and writer Paula Gunn Allen and Jace Weaver, who focus on the power and ethics of Native perspectives, traditions, communities, lands, histories, and sovereignty rights in literary and performance studies[5]. Informing the dramaturgy of *Out of Dust*, Gunn Allen's chapter "The Sacred Hoop: A Contemporary Perspective" offers a particularly helpful analysis of the sacred interrelationship of Native views of land, nature, and community (*Sacred* 54-75). This essay, also reprinted in Cheryll Glotfelty and Harold Fromm's *The Ecocriticism Reader*, opens pluralist theoretical spaces in ecocriticism that move beyond the historic dualism of human and nature in most western traditions. Instead, according to Gunn Allen, Native peoples, while having a multiplicity of traditions and histories, tend to focus on holism: "The notion that nature is somehow over there while humanity is over here or that a great hierarchical ladder of being exists on which ground and trees occupy a very low rung, animals a slightly higher one, and man (and never woman)—especially 'civilized' man—a very high one indeed is antithetical to tribal thought"

---

3 See Weaver, *That the People Might Live* (95-103); foreword to *The Cherokee Night and Other Plays* (xi-xv), and *Notes from the Miner's Canary* (316-31).

4 For Weaver's analysis, see note 2. Stanlake provides an insightful Native platial analysis of *The Cherokee Night* in *Native American Drama* (42-60, 71-77, 87-93). For other critical analyses, see also Darby, "Broadway (Un)Bound"; Driskill, "Ha'nts"; Justice, *Our Fire Survives the Storm* (92-104); and Womack, *Red on Red* (271-303). For a thoughtful analysis of Riggs's plays in terms of his relationship to the land, including *Out of Dust*, see Cox, *Red Land* (65-105).

5 For further elaboration, see Gunn Allen, *The Sacred Hoop* and *Off the Reservation* (esp. 132-78); and Weaver, "More Light Than Heat," *That the People Might Live*, *Notes from a Miner's Canary*, *Other Words*, and his edited volume, *Defending Mother Earth*. See also Geiogamah and Darby, eds. *American Indian Performing Arts*.

(*Sacred* 59). Emphasizing these spiritual interconnections, she elaborates an understanding of "all creatures as relatives (and in tribal systems relationship is central), as offspring of the Great Mystery, as co-creators, as children of our mother, and as necessary parts of an ordered, balanced, and living whole" (*Sacred* 59). She further explains this ecological perspective "assumes that the earth is alive in the same sense human beings are alive," rather than perceiving land as inanimate as in many western perspectives (*Sacred* 70).

This belief that all creation is interrelated requires responsible, respectful, and reciprocal ethics from humans to maintain balance and harmony. Gunn Allen describes this sacred ecology: "For tribal peoples, spirituality and mysticism are communitarian realities. The community and every individual within it must ever be mindful of the human obligation to spirit, balance, and the relationship (or kinship) that exists among all beings, so that all might prosper" (*Off* 47). Jace Weaver also stresses the "linkage of land and people within the concept of community" (*That* 38). Consequently, he notes: "Natives traditionally do not relate to the land as landscape", explaining the word's etymology "shares a common origin with the injunction of Genesis 1:28 to have dominion over the creation" (*Other* 302). Rather than domination, Weaver clarifies that human responsibility "involves not only right relation between the human self and human others but between self and place" (*Other* 35). European invasion into Native communities, on the other hand, brought epistemologies of nature as consumable resources and legal constructs of land as private property that were radically different and historically destructive to Native nations and their communal lands. Environmental historian Sarah T. Phillips explains that "the frenetic pace at which the continent was conquered and its native inhabitants subdued, combined with the relentless drive to submit land, woods, water, and wildlife to the discipline of profit-hungry markets—all this altered ecological conditions and created new forms of power and authority" (287).

Complementing Paula Gunn Allen's and Jace Weaver's work is Theresa J. May's theoretical move in her aptly titled article, "Beyond Bambi: Toward a Dangerous Ecocriticism in Theater Studies," May, a scholar, director, and ecodramaturg, stresses the complex interplay of performance, human beings, nature, and broader environmental issues. Building on Una Chaudhuri's foundational work, she urges "interdisciplinarity" in theater and performance studies "to engage not only literary ecocriticism, but perhaps more importantly, cultural studies and environmental history as well" (97). Acknowledging the unique human interactions afforded by performance, rather than analyzing nature in isolation, she further explains: "A dangerous ecocriticism will participate in the discourse of environmental justice by foregrounding the body as medium between material and metaphoric worlds, and mapping the connections between social injustice, human and other bodies, and environmental exploitation" (101). Integral to "dangerous ecocriticism," according to May, is also broadening the discourse to interrogate perspectives, ideologies, and power relations in terms of lived realities. A central aspect of this approach is recognizing that "[h]egemonic values are often held in place by iconography and long-ingrained master narratives, while the associated devastation, genocide, ecocide, or injustice is occluded" (100), and then contesting them. May concludes with an ecoactivist vision for performance: "At a time when so many longstanding human endeavors seem to have endangered the integrity of all earth's species, ecocriticism can expose and dismantle the ideologies and master narratives that shore up business as usual and inspire performances that help society recalibrate toward sustainable human means" (104). With Wendy Arons, May advances this perspective in *Readings in Performance and Ecology*, an edited collection focusing on the interplay of performance, ecocriticism, and environmental activism. Her views add urgency to Lynn Riggs's "Credo" over half a century earlier.

Contested Views of Civilization

In terms of production values, *Out of Dust*, produced by the Theatre Guild at the Westport Country Playhouse in 1949, directed by Mary Hunter with scene design by Robert Edmond Jones, was positioned for commercial success with a scheduled Broadway opening in October. Originally written "in verse," and then revised "in prose for the Guild production," *Out of Dust* retained its "poetic" (85) quality according to Charles Edward Aughtry (85). Even the program proudly proclaimed, "PRIOR TO BROADWAY!"[6] Instead of its promised Broadway run, however, the play only had a short professional tryout in Westport from August 8-13. In discussing this decision, Aughtry noted: "It was good enough to engage the interest of a highly talented director, designer, and cast (to say nothing of the Guild itself), but, it was not thought to be a good financial bet and so fell into the limbo that several of Riggs's plays unjustly occupy" (133-34). In 1950, Riggs reworked the play, according to Aughtry, in hopes of a future production (85).

Yet, from an ideological perspective with its tragic counter to the often triumphal American West of recent Broadway productions, the play's lack of commercial viability is not altogether surprising.[7] *Out of Dust*

6 For discussions about the Westport Country Playhouse, Laurence Langner, Armina Marshall, and Theresa Helburn and their impact on new play development, see Langner, *The Magic Curtain*, and Somerset-Ward, *An American Theatre*.

7 See Weaver, "Ethnic Cleansing, Homestyle."

reverberates with betrayal as Teece, Bud, and Jeff Grant, three brothers on the Shawnee Cattle Trail, driven by their cruel, patriarchal father Old Man Grant, and manipulated by sinister, former convict King, plot their father's murder. The opening night review in *Variety*, while extolling Robert Edmond Jones's "series of prairie views" as "quite breathtaking," soundly dismissed the play. The critic complained, "Lynn Riggs is in a somber mood" and insisted the play "hasn't much for the commercial theatre" and is "pretty heavy theatre going" (Doul 58). Yet, for me, the "somber mood" is actually its power, for *Out of Dust* remains a pivotal play in Native and United States theatre, probing the moral tensions of nation-building discourses used to legitimate breaking up Indian Territory, seizing Native lands, and dispossessing their communities at the end of the nineteenth century.

While scholarship on federal Indian policy in the late 1800s is multifaceted, a brief look at national narratives about the American West and the history of injustices against Native peoples will help inform Riggs's dramaturgical choices in *Out of Dust*. Drawing on the historical work of Richard Slotkin and Patricia Nelson Limerick, Armando José Prats argues the "Myth of Conquest" (82) dominating popular media, especially the genre of the Western, served the larger ideological function to "define, highlight, and extol Conquest's own values and virtues" (85). In terms of obfuscating Native histories and sovereignty rights in the United States, these masked power relations fueled a mythology justifying European American conquest of Native western lands and the so-called triumphal rise of civilization.[8] Animating these narratives was an oppositional dichotomy of "civilization" and "savagery," based on colonial constructions deeply engrained within European American ideologies and Indian-White political relationships.[9] What is helpful in considering the ethics of *Out of Dust* is the question of how policy makers in the late 1800s leveraged a highly problematic European American construction of civilization conflating patriarchy, private property, capitalism, and Christianity. This campaign helped dispossess the Cherokee people and other nations in Indian Territory and pass the General Allotment Act of 1887 and the subsequent Curtis Act in 1898, federal legislation which

8 This is a major field, which deserves much more discussion than the brief mention provided. Notable works include Deloria and Salisbury, eds., *A Companion to American Indian History*; Deloria, Jr., *Behind the Trail of Broken Treaties*; Hine and Faragher, *The American West*; Limerick, *The Legacy of Conquest* and *Something in the Soil*; and Slotkin's trilogy.

9 For a detailed discussion of European and European American constructions of "Indianness," especially in popular culture and media, see Berkhofer, Jr., *The White Man's Indian*; Deloria, *Playing Indian*; Prats, *Invisible Natives*; Weaver, "Ethnic Cleansing, Homestyle"; and Wilmeth, "Noble or Ruthless Savage?".

opened the way for Oklahoma statehood.[10] Emphasizing that "the conduit of this policy was land," Christy Stanlake writes: "The objective of the General Allotment Act was twofold: first, it aimed to transfer more Native American lands to non-Native peoples; second, it worked to further weaken Native communities by dissolving their cultural networks from within the tribally held lands" (35). The results broke Indian Territory into allotments of individual parcels of 160 acres for each head of household, stripped sovereignty rights from nations, and seized millions of acres of land later sold to European American settlers for homesteading and farming. According to Stanlake, "the 138 million acres of lands reserved for Native Americans shrank to 52 million acres" (36). In their research, Vine Deloria, Jr. and Wilcomb Washburn recount the devastating effects of this loss of well over half of treaty-protected tribal lands in Indian Territory and attacks on Native spiritual, social, cultural, and political systems. Daniel Heath Justice maintains "the General Allotment Act of 1887 stands as one of the most traumatic pieces of U.S. legislation ever imposed upon the Cherokees, second only to the Indian Removal Act of 1830" (91).

These federal policies, while clearly in violation of earlier treaties, intentionally sought to disrupt the integral connections between land and kinship relations central to communities. As previously discussed, Native communities, while richly diverse, according to Paula Gunn Allen, "view the land as holy—as intelligent, mystically powerful, and infused with supernatural vitality" (*Off* 41) in terms of relationships and interconnected kinship ties. This spiritual perspective on land directly contradicted European American perspectives on nature and private ownership of land. Jace Weaver clarifies the ecological implications of this distinction: "Such a view stands in stark contrast to the myths of conquest and dominance that turn the land, which was once community, into mere property and the other-than-human persons and Mother Earth herself into simple commodities to be exploited" (*That* 163).

During the allotment period, leading up to Oklahoma statehood, attacks on Indian Nations' sovereignty, lands, and communal traditions were often

10 For a historical analysis on Federal Indian policy and the ramifications of allotment and later Oklahoma statehood, see Conley, *The Cherokee Nation* (179-205); Debo, *And Still The Waters Run*; Denson, *Demanding the Cherokee Nation* (1-13, 201-42); Deloria, Jr., *Behind the Trial of Broken Treaties,* especially (6-11); Hagan, *Taking Indian Lands*; Holm, "Indian Lobbyists"; Otis, *The Dawes Act and the Allotment of Indian Lands*; and Washburn, *The Assault on Indian Tribalism*. For insightful analysis of the effects of allotment policies on Cherokee families, see Stremlau, *Sustaining the Cherokee Family* and "To Domesticate and Civilize." For Native resurgence in Oklahoma in the twenty and twenty-first century, see Clark, *Indian Tribes of Oklahoma*, and Conley, *The Cherokee Nation* (213-42).

justified under the banner of advancing "civilization," a construction Riggs contests throughout *Out of Dust*. A. B. Meacham, former Superintendent of Indian Affairs, in his chilling 1880 editorial, "Our Red Brethren—Extermination or Civilization—Which Shall it Be?" in *The Council Fire* clearly articulated this position. In a colonizing stance, he turned to his fellow European Americans: "Let us as the stronger race accept the responsibility as becomes a Christian nation. There stands the problem. Which shall it be, civilization or extermination?" (1) As Helen M. Bannan, who also cites Meacham, further articulates, embedded in this notion of "a Christian nation" was an essentialist view of civilization value laden against Native lifestyles, beliefs, and communal lands. Clarifying this construction, Bannan writes: "Civilization simultaneously referred to: 1) the process of transformation by which 'savage' or 'barbarous' Indians would become 'civilized' Americans; and 2) the end product of that process, an ideal state of social organization, existing in its most highly developed form among the white Anglo-Saxon Protestants of the United States" (788). Central to this process, according to Rose Stremlau, were the division of Native lands into individual lots, the promotion of private ownership, the disruption of communal relationships and extended kinship patterns, especially matrilineal ones, and "turn[ing] to federal Indian policy to fracture these extended indigenous families into male-dominant, nuclear families, modeled after middle-class, Anglo-American households" ("To" 265). Thus, rupturing tribal ties and revoking sovereignty rights were integral to allotment plans for "civilizing."

The so-called "Friends of the Indian" were a group of Protestant reformers who met annually at Lake Mohonk Conferences in Massachusetts in the 1880s. Affluent and politically connected, they were particularly successful in shaping and in many ways sanctioning national policy leading up to allotment.[11] Among these, the Reverend Lyman Abbott, a leading Congregationist minister, was particularly influential. He interwove an expanded view of Darwin's theory of evolution with his interpretation of Christianity.[12] According to his biographer Ira V. Brown, Abbott "went on to apply the evolutionary principle to the Bible, theology, the church, society, and the soul" ("Lyman Abbott" 223). In a seeming contradiction with Christ's teachings about the poor, his social evolutionism went hand in hand with Americanization, the rise of industry, and the acquisition of land. Thus, while drawing on the New Testament, Abbott interpreted individualism, property ownership, and capitalism as

[11] For an in-depth analysis, see Bannan, "The Idea of Civilization," and Maddox, *Citizen Indians*.

[12] See also Abbott's *The Evolution of Christianity*, and Brown's biography, *Lyman Abbott, Christian Evolutionist*.

inherently Christian values.[13] As Lucy Maddox further articulates: "At the heart of Abbott's conception of a Christian civilization is the idea of individual self-sufficiency and ambition, an idea he and other Protestant reformers saw as completely antithetical to traditional tribal practices" (80). Maddox continues that Abbott and others maintained "the problem was that the Indians' aversion to capitalist competition (at least as the reformers saw it) was clearly anti-Christian" (80). Thus, these European American reformers failed or refused to recognize the reciprocal spiritual and social power of Native communal views of land and kinship ties.

Therefore, fueling this European American view of civilization, as Angie Debo explains, was "an almost mystical faith in the value of private ownership [of land]" (21) and "theoretical belief in the sanctity of private ownership" (22). At the heart of this national debate was the commodification of the Earth as a marker of civilization. In an influential address made to the 1885 meeting at Lake Mohonk, Senator Henry Dawes from Massachusetts used the highly successful Cherokee Nation in Indian Territory, having rebuilt after the devastation of the Trail of Tears, as his example of failure. He discussed the strengths and prosperity of the Cherokee Nation, where every family had a home and "not a pauper in that Nation," even noting "its schools and its hospitals." In spite of this compelling evidence, Dawes still insisted: "Yet the defect of the system was apparent. They have got as far as they can go, because they own their land in common" (43). For Dawes their communal harmony, social responsibility, and shared lands, rather than showcasing their advancement, revealed deficiency. Thus, he concluded: "There is no selfishness, which is at the bottom of civilization. Till this people will consent to give up their lands, and divide them among their citizens so that each can own the land he cultivates, they will not make much more progress" (43). Espousing selfishness as a virtue, Dawes condemned the Cherokee people for their lack of greed to pursue capitalism and hold private property, instead of supporting their social commitments to all community members and respect for the land as hallmarks of civilization.

As politics of power, Senator Dawes's ideology of civilization privileging individual selfishness, greed, and land ownership as national values had far-reaching ramifications. In the U.S. Senate, he sponsored the General Allotment Act of 1887, often called the Dawes Act, which became federal law and helped create the new state of Oklahoma. This act also strengthened the capitalist push of European American economic forces, propelled by what D. S. Otis called "land-hungry western settlers" (20). Donald L. Fixico stresses the connections between "exploitation in the

[13] For a more articulated discussion of Abbott's views about Christianity and capitalism, see *Christianity and Social Problems.*

form of American capitalism" and "desecrating the earth and destroying the spiritual and physical harmony and balance of all things" (41). Environmental historian Donald Worster also draws a clear link between aggressive capitalist ideology and environmental destruction of Native lands: "When the white men came to the plains, they talked expansively of 'busting' and 'breaking the land'. And that is exactly what they did" (4). For Worster, the extensive ecological damage of the Dust Bowl, massive dust storms in Oklahoma and surrounding states throughout the 1930s, "was the inevitable outcome of a culture that deliberately, self-consciously, set itself that task of dominating and exploiting the land for all it was worth" (4).

Daniel Heath Justice, revealing the ethical incongruity of allotment policy, "ostensibly intended to bring Indians to the level of civilization enjoyed by the U.S. citizenry," contends: "The irony was, of course, that the autonomous republics of the Five Civilized Tribes, especially the Cherokees, had long exceeded the 'civilization' of their White neighbors" (89-90). Thus, the Cherokee Nation rigorously opposed the General Allotment Act and later the Curtis Act.[14] In defending Cherokee Nationhood and lands, according to Andrew Denson, these activists were astutely aware of the "nature of American Indian policy" in these arguments centering on "civilization." Consequently, these leaders "picked apart American efforts to disavow racism and oppression and insisted that if the Cherokees were to be dispossessed of their government and property it would be a crime and not the working out of some universal principle" (10). So profound were the effects on Riggs, Justice asserts: "The allotment dispossession of the Cherokees of Indian Territory clearly shaped the course of Lynn Riggs's life, just as it continues to impact many Cherokees today and our relationships to our communities, tribal connections, and cultural responsibilities" (108). Sharing similar views, James H. Cox also notes the impact of Oklahoma on the playwright: "Indeed, Oklahoma statehood for Riggs stands as not only a disruption of Cherokee sovereignty but also an attendant disruption of traditional land use in the late nineteenth century in Indian Territory" (104-5). The dramaturgy of *Out of Dust* suggests that Riggs continued Cherokee activism, contesting the politically expedient discourses around "civilization" and the exploitation of Native lands as spiritually corrupt, inherently barbarous, and ultimately tragic.

A Crucible of Power

Set in Indian Territory, *Out of Dust*, is geographically, culturally, spiritually, and ethically rooted in Native lands and traditions before allotment and Oklahoma statehood. Cynthia Fowler describes diversity in

---

[14] See especially Holm, "Indian Lobbyists," and Hoxie, ed. *Talking Back to Civilization.*

Indian Territory and later Oklahoma: "the home of Cherokees, Choctaws, Chickasaws, Creek, and Seminoles who formally lived in the East. By the mid-nineteenth century, Kiowas, Kiowa-Apaches, Wichitas, and Osages had also moved to this area" (10). Locating European Americans "on the Shawnee Cattle Trail through Indian Territory, in the early 'eighties'" (215), Riggs also signals white intrusion and exploitation. Once a Native trail, Wayne Gard recounts the trail's more recent history with "the long line of settlers' wagons [...] military supply caravans and companies of freighters. Many of the freighters carried buffalo robes or other pelts on their return north" (262). Hence, these activities were complicitous with aggressive slaughtering of almost "20 million" (Worster 77) buffalo on the plains that nearly drove them to extinction. On the way to cattle markets in the 1800s, this trail cut through the Choctaw Nation and later through the Cherokee Nation (261-62). The play's location also dramaturgically and geographically ties the performance to these nations before Oklahoma statehood.

From a dramaturgical perspective that considers performance elements as well as the script, the choice of Robert Edmond Jones, a leading innovator, as the set designer suggests the geographic impact Lynn Riggs wanted the prairies of Indian Territory to have throughout the entire play. Land in Native theatre cannot be overemphasized, according to Christy Stanlake, who maintains: "Place figures so predominantly in Native American plays that one could argue that it carries the same value as characters do and often functions just as actively" (39). In terms of performance, she continues: "Staging this awareness of the land's profound meanings is one of Native theatre's strongest dramaturgical attributes" (39). Thus, Jones's work was central to staging the first production of *Out of Dust* at the Westport Country Playhouse in August 1949, and his sets according to the *Variety* critic were the performance's power (Doul 58). A friend and colleague of Riggs, Jones had designed the "settings" for the 1930 Broadway production of *Roadside* at the Longacre Theatre (Riggs 1). Supportive of Riggs's Cherokee background, he had also endorsed the publication of *Russet Mantle and The Cherokee Night* in 1936, writing: "Cherokee Night is one of the finest plays yet written by an American" (dustcover). According to Ralph Pendleton, Jones's design artistry was particularly evocative for land and space.

In *The Dramatic Imagination*, Jones's reflections on his theory of stage design also offer insights into the ecological impact of the sets in *Out of Dust*. Jones explains: "The task of the stage designer is to search for all sorts of new and direct and unhackneyed ways whereby he may establish the *sense of place*" and "*to remind the audience of where the actors are supposed to be*" (135). For him then, "A true stage-setting is an invocation to the *genius loci*—a gesture 'enforcing us to this place'"

(135-36). Thus, he cautions against explicit realism: "A good scene, I repeat, is not a picture. It is something seen, but it is something conveyed as well: a feeling, an evocation. [...] It is a presence, a mood, a warm wind fanning the drama to flame. It echoes, it enhances, it animates" (26). Jones's close collaboration with Riggs on the 1949 production and the stage directions suggest that the prairies of the Southern Plains served to evoke Indian Territory as a watchful presence. Scene after scene, they also connected the play to vast natural grasslands before the devastating effects of European American settlement in Oklahoma and nearby states. According to Donald Worster, this was a time when "the grass still rippled under the wind" (78). Soon damaging plowing, extensive erosion, and aggressive wheat farming "destroy[ed] the delicate ecological balance" and contributed to the Dust Bowl (Worster 5). Jones's striking sets offered a powerful contrast to harrowing news reports and devastating images of dust storms, drought, and abandoned homes that documented these struggles throughout the 1930s.

Set against the open lands of Indian Territory and their resonance with Cherokee life before allotment and Oklahoma statehood, Riggs probes ethical dangers facing this European American family through a mastery of form to match his narrative. Eloise Wilson, a Cherokee who knew Riggs personally, noted: "In 'Out of Dust' three great influences upon Riggs—the Bible, Shakespeare, and folk song—have been compounded into a three-dimensional drama" (34). These are framed by his Cherokee background and commitment to community. Rather than a political polemic, Riggs as a Native playwright deftly interweaves these elements and constructs the play as a frontier morality tale. The taunt structure helps unmask the underlying moral contradictions of the mainstream triumph of civilization in the American West often portrayed in popular culture. The patriarchial father Old Man Grant drives his sons to a breaking point when they turn on him and plot his murder. Intertheatricality with William Shakespeare's *King Lear* invokes the evil inherent in the breakdown of kinship relationships within European contexts and the dangers of placing lust for individual property over family. As Wilson further explains, "Riggs's knowledge and love of Shakespeare is evident: Old Man Grant is the King Lear of the Shawnee Trail, denied by and plotted against by his older sons but loved by the youngest" (34).[15] By dramatizing these devastating consequences, this story is not simply about one European American family, but an interrogation of the lived dangers of ideologies fueling Westward expansion into Native lands.

[15] See also Braunlich, who concurs with Wilson's analysis, in "The Oklahoma Plays" (393), and Weaver, foreword to *The Cherokee Night and Other Plays* (xiv), and *Notes from a Miner's Canary* (325).

Souls hang in the balance as the play opens. Cloaked in darkness, the performance begins in a shack, a site of alienation and disconnection from the surrounding plains. Evoking transgressions, a distraught Maudie Grant, dimly lit, sets a penitential tone by beginning to confess to a mysterious Mr. Osborn. Warning that "what happened to us on the Trail could happen to anyone", Maudie sets the moral dimensions of the play around human "weaknesses…temptations…greeds" (216). The structural use of confession for events "on the Shawnee Trail" also subtly undermines European American Western expansion with connections to sin and damnation (217). Maudie begins her story; then lights go up and a flashback introduces the destruction that brought her to this desolation.

In the next scene, Riggs's stage directions initially suggest the popular iconic setting of the American Westward movement—cowboys, cattle trails, and covered wagons:

> *Two* COWBOYS *are singing as they draw nearer. The light lifts and widens. It is sundown in a camp on the Old Shawnee Trail to Baxter Springs through the Indian Territory in the early 'eighties. The end of a covered wagon can be seen—and some accoutrements of travel. Scrubby bushes and a lone tree and deep space to the horizon.* (217)

However, in this theatrical space, geographic, physical, and spiritual conditions of the trail soon complicate the familiar semiotics of the popular American West. Far different from romanticized cowboys, Luke and Snake joke about their exhaustion. Then Jeff, the youngest son of Grant, describes the "dust" of the cattle trail, especially "ridin' drag" (221) behind the herd: "Like somethin' snatched out of hell! It's a brute's job back there. […] The dust mile high from the hooves ahead. Till you can't see. And you can't breathe. And after a while you don't want to. You want to lay down and die" (222). Rather than connecting to the land as a site of continuity and spirituality, as in the Cherokee tradition, this family's connection is to dust and death. The imagery of a "mile high" dust cloud caused by hundreds of cattle also has important environmental implications undercutting the drive's legitimacy on the Old Shawnee Trail. Cattle's destructive over-grazing of natural grasses "was a foretaste of later developments on the plains" (Worster 83) leading to the ecological tragedy of the Dust Bowl in the 1930s.

The reference to "dust" has further religious implications. The word resonates with the title "Out of Dust" and evokes the Judeo-Christian creation story in Gen. 2:7 (King James Version): "And the LORD God formed man *of* the dust of the ground, and breathed into his nostrils the breath of life; and man became a living soul", and subsequent temptation, fall of humanity, and expulsion from Eden. The name Snake, and reference to "hell" conjure up punishment of the serpent in Gen. 3.14 and

Isa. 65.25, furthering the transgressive nature of the ranchers' cattle drive through Indian Territory.

Dramaturgically, the first act of *Out of Dust* lays bare moral fissures in this nuclear rancher family, headed by the father Old Man Grant. The stage directions describe Grant as "*patriarchal, and severe, bigoted, tyrannical and deeply—mystically—a troubled man. His search has carried him through hard-won accomplishment, through greed and lechery and power over men. He has dignity and authority, dimensions that are hard to deal with*" (229). Eloise Wilson first noted the biographical resemblance between Riggs's father and Grant (118), an interpretation Jace Weaver more fully elaborated.[16] A biographical reading is certainly valuable. However, the dramaturgy of *Out of Dust* also suggests that Riggs in Grant, as the patriarch, constructs a crucible of power that inexorably leads to the man's own destruction and that of his family. A key component of patriarchy, as S. Scott Bartchey explains, is not just "male domination of females" (136). Rather, he elaborates: "In a wide variety of cultures, men are brought up to gain honor for themselves precisely by dominating as many others as they can, both men and women" (136). As discussed earlier, promoting patriarchy through nuclear families was also a major strand of allotment policy, according to Rose Stremlau, because "reformers concluded that kinship systems, especially as they manifested in gender roles, prevented acculturation by undermining individualism and social order" ("To" 265). Countering this view, the staging in *Out of Dust* enacts the Grant nuclear family's increased dysfunction and suffering under the weight of the father's power.

Act 1 dramatizes Grant's hold over his family and establishes his destructive role in the plains. His youngest son Jeff first introduces him as the trail boss of his cattle herd, not as a parent: "He can roam all over—and no one can say Boo to him" (222). His role as a successful cattleman is dramaturgically and ecologically significant. "The world of cowboys, roundups, and cattle drives" has a far different "ecological story" (83), according to Donald Worster, than popular perceptions. He writes that "the cattlemen pushed the land as far as it would go, and then pushed some more. They generally viewed the southern plains as another Comstock Lode, to be mined as thoroughly as possible by overstocking the range" (83). Stressing their exploitation of the land, Worster documents the destructive impact of over-grazing: "In some areas they ran four times as many cattle as the grass could carry, resulting in depletion and long-lasting damage. By 1880 a steer on one range needed 50 acres to fatten up, when a decade later 5 acres had been sufficient" (83).

16 Weaver further elaborates this connection in the foreword to *The Cherokee Night* (xiv), and *Notes from a Miner's Canary* (327).

Disconnected from his family and embodying the capitalist ideology that also marked mainstream conceptions of civilization, Grant has put his cattle business and financial bottom-line over his relationship with his sons, and hired an outsider, King, an ex-convict, as second-in-command of his herd. The audience also learns that Maudie, now married to Teece, endured Grant's sexual abuse as a young orphaned caregiver for his sons. Caught in his father's grip, Jeff longs for an alternative to life on the cattle trail and his father's pursuit of money, insisting, "I just want to live" (222). Characterized by "open-heartedness" in contrast to his father, Jeff yearns for a family and happiness (Wilson 34). Similar to Curly in *Green Grow the Lilacs*, he bursts with exuberance over the prospect of seeing his fiancée Rose Dunham soon and his joy over their impending marriage, a plan his father resists. Thus, the cattle drive also functions metaphorically on a number of levels for Grant's treatment of his family—his sons, Teece, Bud, and Jeff, and daughter-in-law, Maudie—and the larger encroachment on western Native lands and subsequent environmental damage to the southern plains.

As the play unfolds, family relations falter under the abuse of Grant's power and ruthless individualism, pillars of U.S. nation building. Angered that his father hired King as trail "sub-boss" instead of him, Teece insists on more responsibilities and higher wages than his current "cowhand's job" (230). Smug in his sense of entitlement, echoing social evolutionism, Grant counters his son's request with his credo of survival of the strongest: "Want, want! Things in this world are for men who take 'em" (230). When Teece counters, Grant cruelly disparages his manhood: "A man would see his way clear to get 'em, anyway. Without their bein' give to him" (231). Rather than acknowledging his kinship obligations as a father to his son, he blames Teece for his current situation, taunting him that it resulted because he "couldn't take charge." He reminds him: "You have to grab things when they come your way—and hold on to 'em" (231). Putting his business before family, Grant further alienates his eldest son by reminding him that he "*signed a paper*" (231) and is legally bound to continue working for him as his employee. He also refuses to pay Bud any salary until he works off money owed his father. When all three sons want to rest after the exhausting cattle drive, Grant refuses. Lured by even greater profits, he orders his sons immediately to head "home [...] to round up another herd," insisting "that's what we'll do. Start on the Trail again" (232). Recognizing the larger historical and ecological implications of this scene, Jace Weaver in his insightful analysis of the play observes that "Grant is speaking more broadly, his attitude that of avaricious Anglo Oklahomans about the taking of Native lands" (*Notes* 325). This scene also denaturalizes Grant's patriarchal power through his insistence on the contractual nature of property over family ties, in sharp contrast to communal values, and his commodification of the animals he is herding.

Dramaturgically, Riggs's construction of Grant's corruption of Biblical teachings also subverts the trail boss's moral authority and ideology. Seeking to justify his domination after he arrived in "the Territory," Grant explains to Teece: "The men in the Bible wrestled with the desert and made it bloom. They stood alone at the head of their family, at the head of their flocks and fields, and asked no favors. And expected none. I read about 'em. Night and day, they was close to my thoughts" (231). Grant's reading of the Old Testament suggests an appropriation of scripture to justify his own credo of taking. This perspective, according to John Mohawk, is a form of "anthropocentrism—which comes directly from the [European] tradition calling upon 'man' to go out and assert 'dominion' over nature" (442). Through Grant's conflation of God and worldly power, this scene also resonates with positions of Protestant reformers, especially Reverend Lyman Abbott. In his influential essay "Our Indian Problem," Abbott invoked Christianity to sanction "pioneer" rights to treaty-protected lands in Indian Territory. With anthropocentric certainty, he argued:

> The tide of civilization, surging westward, comes some day to a fair and wealthy but unused and idle territory. There are forests which no woodman's axe has ever touched; rivers where waterfalls turn no mill wheels; mountains whose treasures of gold or silver, iron or copper or coal no pickaxe has uncovered; prairies whose fertile soil is prolific only in weeds. "Come," cries the pioneer, eager to develop this useless territory, "let us go in and make those acres rich with our industry." [...] "When are they [Indians] going to use this land; to convert this timber into boards; these rivers into mill-streams; when are they going to excavate these minerals and turn these weedy prairies into fruitful farms?" (725)

In a chilling rationale that is painful to repeat, Abbott justified European American aggression with divine law: "Barbarism has no rights which civilization is bound to respect" (725). With ruthless racism, he then insisted: "Let us understand once and for all that an inferior race must either adapt and conform itself to the higher civilization, wherever the two come in conflict, or else die. This is the law of God, from which there is no appeal" (727). Implicit in these remarks, as in Grant's dialogue with his son Teece, is a breakdown of spirituality and ethics that has profound ecological impact. According to Linda Hogan, Chickasaw writer, "a language of commerce and trade, of laws that can be bent in order that treaties might be broken, land wounded beyond healing" fails to honor the sacredness of relationships among all of creation (45-46).

At the end of scene 1, Maudie confronts Grant's distortions and offers her own interpretation of scripture, intensifying the spiritual stakes. Still a victim of his control, she fends off his unwelcome sexual advances, telling him that she always hated him for his violation of her as a young woman and mistreatment of his son Teece. Her speech anticipates the rest of the performance when she reminds him: "Sins have to be answered for, Mr. Grant. Mine will. Yours will, too. You'll grovel in the dirt someday—like I do now. It's comin' to you" (239). When he refuses to show remorse, Maudie cites Jer. 5.6, further warning him of the spiritual consequences of his actions: "You know it's so. It's in the Bible. '… a lion out of the forest shall slay them … a leopard shall watch over their cities, everyone that goeth out thence shall be torn to pieces… …'" (239). Implied in Maudie's speech is the rest of the verse "because their transgressions are many, *and* their backslidings are increased" (Jer. 5.6 [emphasis in the original]). Thus, she personally indicts his credo of taking as it applies to his "taking" of her in her youth and his brutality to his sons. The scene further impugns Grant's morality as he resorts to another distortion of the Bible to justify his transgressions. Secure in his privilege and power, he asserts: "No. Some that walks the earth is above the laws laid down for the weak. '… a lion out of the forest shall slay only them that oppose us'" (239). In this spiritual contest, Maudie's dialogue refutes Grant's view as a moral corruption of the New Testament, insisting: "You've twisted the meaning. '… the wages … is death'" (240). This reference to Rom. 6.23 is a powerful close to the scene. As a broader condemnation, Maudie's Biblical reference to death—both physical and spiritual—seems to undermine the theological underpinnings of both Grant's personal beliefs and a perspective of civilization that conflated Christianity, worldly power, violence, and dominion over nature.

Later in juxtaposition, in a key scene with Jeff, Maudie offers an invocation resonant with kinship values and love, based on Paul's teachings about Christian fellowship in the New Testament. Wanting to marry Rose and settle down, Jeff is furious that his father has ordered him back on the Trail. Maudie, fearing Jeff is becoming hardened like his father, urges him to move from rage to harmony: "Think of other things. Singin', like we done just now. People laughin' and happy. With grub to eat. Love by their side, like you've got now." Drawing on Phil. 4.8, she concludes: "'Think of these things.' The Lord said to" (276). The full verse is illustrative of lived values: "Finally, brethren, whatsoever things are true, whatsoever things *are* honest, whatsoever things *are* just, whatsoever things *are* pure, whatsoever things *are* lovely, whatsoever things *are* of good report; if *there be* any virtue, and if *there be* any praise, think on these things" (Phil. 4.8 [emphasis in the original]). By invoking Philippians, on one level, Maudie's dialogue impeaches extreme individualism and selfishness advocated by the father Grant as a violation of the New Testament's broader conception of family and community. S. Scott Bartchey's work on Paul and the early Christian community helps open the dramaturgy in this scene. Bartchey writes that "Paul's apparent goal was not the creation of an egalitarian community in the political sense, but a *well-functioning family* in the kinship sense. In this family, each surrogate member used his or her strengths, whatever they were, to enrich the quality of life for the family rather than for themselves as individuals" (146 [emphasis in the original]). By providing an alternative to Grant's philosophy of taking with this popular Biblical reference, the scene reconfigures ethical parameters for civilized behavior in the play, based on community, parameters the Grant men, King, and even Maudie are unable to sustain.

Rather than creating relationships based on familial responsibilities and love, Grant's hardness and greed inevitably leads to the destruction of his family as he sets events in motion that further propel the performance's tragic trajectory. Here narrative consequences grow organically out of beliefs and actions. Out of touch with the land and even his own horse, Grant falls and is left on the ground until Bud and King find him. Proud of his horsemanship, rather than accepting the fall as an accident, Grant accuses his sons of trying to orchestrate his death by deliberately provoking his horse with a stray "heifer" as he "loped down in the gully" (247). Refusing his sons' avowals of innocence, he further implicates them in a twist of the Biblical account of Joseph (Gen. 37.18-24). He tells of three sons who drag and tie down their sheep-herder father and leave him to die. In his rage, he concludes: "It's just a story. Of drought and dust—and betrayal. Like one you'd read in the Bible. When sons conspired against fathers, when brother warred against brother—and nobody won. (*Clearly accusing*) You're not gonna to have it the way you want it! I won't let you!" (250). Combined with his emotional brutality to his sons and Maudie and with King's greed, this accusation with its implicit retribution plants the seeds for Grant's murder. The words "drought and dust" reverberate with the devastation of nearly ten years of the Dust Bowl, once again tying Grant's voracity to the environmental destructiveness of many of the ranchers, farmers, and land speculators in the plains.

Dramaturgically, evil stalks the stage in Act 2, further reversing mainstream conceptions of "savage" and "civilized." With nothing to lose after Grant fires him, King surpasses his boss's credo of ruthlessness. Motivated by the "pile" of money tied up in Grant's cattle business, King preys on the two older sons' vulnerabilities and exploits them for his own financial ends (234). He manipulates Teece's humiliation over his father's demeaning him as a hired hand and his rage at Grant's debasing of Maudie. He then exploits Bud's debt to him for going to prison for cattle theft without revealing Bud's complicity. Through these threats and the lure of their father's money, King pressures the two brothers to murder him by re-creating their father's riding accident with another heifer. In his love for his father and deep sense of morality, Jeff proves more difficult. Ultimately, it is Grant himself who pushes Jeff too far by accusing him of sexual misconduct with Maudie after witnessing his son comforting her. King, finally sensing Jeff wavering, then warns him that his father will force himself on Rose and make her "*take Maudie's place*" (290). Fearing for Rose and furious at his father, Jeff finally accedes. In a powerful dramatic twist, tragedy unfolds when his sons, once chastised for weakness, enact Grant's credo of manhood defined by "men who take 'em" (230) against him and finally do take charge with King by helping cause an accident to take Grant's life.

The next scene breaks soberly in "*a gully*" with an "*eroded bank*" overlooking "*the endless baking prairie*" (290). Against this view, the brutality of murder appears to subvert the sanctioned violence that often accompanied the taking of the West, laying bare the deadly values of conquest. Staged images of erosion and seared land are metaphorically evocative of the men's crumbling morality and geographically significant as another portent of the Dust Bowl, linking human and environmental violence. After the murder, savagery ruptures the family as King immediately seizes control, and Bud and Teece both insist their father deserved to die. After finding his father dead, Jeff feels great remorse, realizing his crime as a son. He reminds his brothers of their filial responsibilities: "Whatever he was—he was a man, our flesh and blood. Who are we to sit and judge him, to say whether he lived or died? Are we God? Look at us! We're not even men. For guilt is on us all that brought this thing to pass!" (303). Sparking further dissension among them, he reveals a folded piece of paper from his father's body, which he cannot read and lets Maudie read, leaving him "everything" (303). The note reveals another secret, "a smudge" (304) of blood, left from King's knife to finish the job that the set-up accident did not. Maudie, once the ethical voice of the family, first stands by Jeff's accusation of King and then weakens when her husband Teece pressures her. King, now embodied ruthlessness, immediately tries to turn the brothers against Jeff, insinuating that Jeff has the most to gain.

Into this maelstrom of violence arrives Rose to surprise Jeff. Full of innocence and hope for her future with Jeff, Rose serves as a moral compass, bringing the audience back to the couple's love story and plans for a family. In her presence, Jeff suffers a spiritual crisis, unable to tell her the whole truth about his complicity in his father's death, allowing her to blame King. Seeking legal alternatives to more violence, Rose urges him to go to Baxter Springs and turn in King, to which he agrees. Betrayals increase when they learn that King, Bud, and Teece took away all the horses as a way to trap and murder Jeff. With Maudie's distraction, they make their escape on Rose's horse that King then shoots and wounds. His brutality combined with the words "damn" and "hell" heighten his association with evil and damnation (314, 315).

The beginning of Act 3 stages a fleeting vision of love and family set against the destructive forces propelling the Grants. Jeff and Rose on the way to Baxter Springs rest at a deserted shack, described in the stage directions as marked by "*ruination and decay*" (320), emblematic of their shattered lives and the larger failure of community to sustain meaningful relationships with one another and the land. The young couple's poignant dialogue extends their loss and yearnings. Rose observes, "Sad kind of end for a house to come to. It maybe was nice one time," and Jeff responds, "They musta hated to leave it" (320). Looking around at the abandoned stove and rocking chair, Rose marks the parallels to their lives: "They coulda been young, like us. Just married" (320). Intensifying the poignancy, Jeff unloads Rose's saddlebags, which contain "fixin's" (320), which she had originally planned to prepare him on their wedding trip and heralds of future family meals. Against the corrosive brutality, in a brief respite together in the shack, they enact the grace of marriage and "home" (321) that they will never have as a couple. Jeff shifts morally in the presence of Rose's goodness and love. Her image of him as "[a] man that holds his head up and looks at you clear and good" (324) forces him to face the depth of his depravity. He then breaks down in contrition and confesses: "It started from King. Spread out from him like a sickness. We listened to him. We agreed to it. The accident. We was all in it" (325). Shocked and repulsed, Rose orders him to go. The stage directions suggest her depth of grief over lost possibilities: "*A purely animal cry of pain escapes her and she bends over, slashed with agony*" (326). Communal ties unravel as all hope for the life with Jeff she had once dreamed of is now lost.

Violence continues to spiral out of control when Teece, Bud, King, and Maudie break in like a vigilante gang, searching for Jeff. Rose, now remorseful over ordering Jeff to leave, tries to protect him by telling them he has been gone "quite a while" (328). Jeff, who waited nearby, enters to defend Rose. Fully embodying Grant's credo of taking, King prepares to kill Jeff and demands guns from Teece and Bud, reminding them: "*I'm* runnin' things, in case you haven't realized" (329). He then chides them: "But for me you'd all hang" (331), insisting Jeff must die to cover up the murder. In contrast to King's sinfulness is Jeff's contrition. Assessing his life at gunpoint, he instantiates the importance of relationships and realizes the highlight of his life was his love for Rose "that was sweet" and "kept me standin' up—with some hope—when all I felt like was to lay down in the dust and die" (332). Remorsefully, he acknowledges his and the brothers' "rottenness," confessing, "I done a great and turrible wrong. I can't blame King too much." He then tells King, "Let me die" (332). In a desperate turn, Rose shoots King as he aims for Jeff, her innocence lost. Moral anxieties intensify as Teece and Bud, seizing upon the opportunist reprieve of King's death, jump to cover up their father's murder by blaming King for his death and threaten Jeff's life if he does not agree.

Flickers of redemption stir at the end of Act 3. In spite of both the temptation to take the expedient path and save himself legally, Jeff ultimately makes the ethical choice, choosing to save himself spiritually. Maudie, moved by Jeff's example, also accepts her guilt in communal terms and mourns her failure to protect "things put in my keeping. Lives. Young lives. Mine once" (336). Jeff and Rose remain resolute against the brothers' and King's threats. Jeff, knowing that he will spend most of his life in prison or hang, leaves to turn himself in and take responsibility for his actions. In a moving moment of love, Rose agrees to go with him and wait for him. Charles Aughtry notes the use of lighting to illuminate the couple's decision: "At the end of the play, Jeff and Rose ride out, 'the morning sun lighting their faces as they head toward Baxter [Springs]' to deliver themselves to justice for their crimes" (133). This brief sunshine serves as a sorrowful consolation for their long-dreamt marriage, family, and future mornings together lost, but their love and commitment to each other still strong. Caught in the savagery of frontier justice and the grasp of a conception of civilization based on domination, they face loss rather than triumph.

The final scene circles back to the opening scene with Maudie. As she prepares to leave with Osborn, she accepts her guilt, telling him, "I'm ready, Officer. Ready and willing" (339). A powerful coda ends the play, interweaving ethical, spiritual, and ecological issues, as Maudie interrogates "the evil" that destroyed the family on the cattle trail. Focusing on human frailties and fears, she concludes: "Yes. Stuck, like we all are—smack between heaven and hell" (339). While Osborn uses the Christian concept of "purgatory" to clarify her meaning, the performance ends with Maudie's focus on lived ethics: "On earth, I mean. In my case—in our case—this was the earth we knowed and walked on" (339). Here she moves the story from the individualism of "my case" to kinship ties of "our case." According to Jace Weaver, "Maudie cryptically ties all that has transpired to the land" (*Notes* 330). Maudie's reference to "the earth" thus on one level evokes Indian Territory, visually dominant throughout the staging. Her words also have resonance with Cherokee communal responsibilities inherent in those bonds. Confirming the centrality of communal values and responsibilities of kinship, Geary Hobson, Cherokee-Quapaw/Chickasaw writer, explains: "Heritage is people; people are the earth; earth is heritage.

By remembering these relationships—to the people, the land, the past—we renew in strength our continuance as a people" (11). Weaver further elucidates these sacred interconnections, explaining, "the Cherokee word *eloh'*, sometimes translated as 'religion,' also means knowledge, history, culture, law—and land. And it means all these things all the time at the same time" (*Notes* 39). Thus, through Maudie's final words, this morality tale concludes by reminding audiences that evil and corruption are choices in human life, often resulting from the tragic breakdown of kinship ties, communal responsibilities, and connections to the land. The ending thereby honors the ethics of Cherokee beliefs and remembers Indian Territory.

As a play written for Broadway, still awaiting production, *Out of Dust* also theatrically invites all audiences, especially European American ones, to look hard at the moral contradictions in a construction of civilization based on individualism, acquisition, ruthlessness, and private gain that has too often dominated United States nation building and culture and envision a transformative one based on care and responsibility for one another and the Earth. Always the visionary, Lynn Riggs concluded "A Credo for The Tributary Theatre" with his commitment to ethics: "I should think we would take the theatre seriously. I should think we would take life seriously" (167). His vision remains vital today. Foregrounding this commitment to all life, Linda Hogan reminds us in *Dwellings*: "We must wonder what of value can ever be spoken from lives that are lived outside of life, without a love or respect for the land and other lives" (132). From Riggs's perspective, theatre opens spaces for these sacred responsibilities.

Works Cited

Abbott, Lyman Reverend. *Christianity and Social Problems*. Boston: Houghton, Mifflin and Company, 1896.

—. *The Evolution of Christianity*. Boston: Houghton, Mifflin and Company, 1892.

—. "Our Indian Problem." *The North American Review* 167. DV (December 1898): 719-28.

Allen, Paula Gunn. *Off the Reservation: Reflections on Boundary-Busting, Border-Crossing Loose Canons*. Boston: Beacon Press, 1998.

—. *The Sacred Hoop: Recovering the Feminine in American Indian Traditions*. 2nd ed. Boston: Beacon Press, 1992.

Arons, Wendy, and Theresa J. May, eds. *Readings in Performance and Ecology*. New York: Palgrave Macmillan, 2012.

Aughtry, Charles Edward. *Lynn Riggs, Dramatist: A Critical Biography*. Ph.D. diss., Brown University, 1959. Ann Arbor: UMI, 1959. 5904327.

Bannan, Helen M. "The Idea of Civilization and American Indian Policy Reformers in the 1880s." *Journal of American Culture* 1.4 (Winter 1978): 787-99.

Bartchey, S. Scott. "Who Shall Be Called Father? Paul of Tarsus between the Jesus Tradition and Patria Potestas." *Biblical Theology Bulletin: A Journal of Bible and Theology* 33 (2003): 135-47.

Berkhofer, Jr., Robert F. *The White Man's Indian*. New York: Vintage Books, 1979.

The Holy Bible Containing the Old and New Testament, King James Version. New York: Grosset and Dunlap, 1931.

Braunlich, Phyllis Cole. *Haunted by Home: The Life and Letters of Lynn Riggs*. Norman: U of Oklahoma P, 1988.

—. "The Oklahoma Plays of R. Lynn Riggs." *World Literature Today* 64.3 (Summer 1990): 390-94.

Brown, Ira V. "Lyman Abbott: Christian Evolutionist." *The New England Quarterly* 23.2 (June 1950): 218-31.

—. *Lyman Abbott, Christian Evolutionist: A Study in Religious Liberalism*. Cambridge, MA: Harvard UP, 1953.

Chaudhuri, Una. *Staging Place: The Geography of Modern Drama*. Theory/Text/Performance Series. Ann Arbor: U of Michigan P, 1995.

—. "'There Must Be a Lot of Fish in That Lake': Toward an Ecological Theater." *Theater* 25.1 (1994): 23-31.

Clark, Blue. *Indian Tribes of Oklahoma: A Guide*. Norman: U of Oklahoma P, 2009.

Conley, Robert J. *The Cherokee Nation: A History*. Albuquerque: U of New Mexico P, 2005.

Cox, James H. *The Red Land to the South: American Indian Writers and Indigenous Mexico*. Minneapolis: U of Minnesota P, 2012.

Darby, Jaye T. "Broadway (Un)Bound: Lynn Riggs's *The Cherokee Night*." *Baylor Journal of Theatre and Performance* 4.1. *Nations Speaking: Indigenous Performances Across the Americas*. Special Issue (Spring 2007): 7-23.

Dawes, Henry L. "Senator Dawes' Address." *Proceedings of the Third Annual Meeting of the Lake Mohonk Conference of Friends of the Indian. October 7 to 9, 1885*. Philadelphia: Sherman and Co., Printers, 1886. 34-45.

Debo, Angie. *And Still the Waters Run*. Princeton: Princeton UP, 1940.

Deloria, Philip J. *Playing Indian*. New Haven, CT: Yale UP, 1998.

— and Neal Salisbury, eds. *A Companion to American Indian History*. Oxford: Blackwell, 2002.

Deloria, Jr., Vine. *Behind the Trial of Broken Treaties: An Indian Declaration of Independence*. New York: Delacorte P, 1974.

Denson, Andrew. *Demanding the Cherokee Nation: Indian Autonomy and American Culture, 1830-1900*. Lincoln: U of Nebraska P, 2004.

Doul. Rev. of "Out of Dust." *Variety* 175 (10 August 1949): 58.

Driskill, Qwo-Li. "Ha'nts: The Booger Dance Rhetorics of Lynn Riggs's *The Cherokee Night*." Ed. Hanay Geiogamah and Jaye T. Darby. *American Indian Performing Arts: Critical Directions*. Los Angeles: UCLA American Indian Studies Center, 2010. 179-96.

Fixico, Donald L. "The Struggle for Our Homes: Indian and White Values and Tribal Lands." Ed. Jace Weaver. *Defending Mother Earth: Native American Perspectives on Environmental Justice*. Maryknoll, NY: Orbis Books, 1996. 29-46.

Gard, Wayne. "The Shawnee Trail." *Southwestern Historical Quarterly* 56.3 (January 1953): 359-77.

Geiogamah, Hanay, and Jaye T. Darby, eds. *American Indian Performing Arts: Critical Directions*. Los Angeles: UCLA American Indian Studies Center, 2010.

Glotfelty, Cheryll, and Harold Fromm, eds. *The Ecocriticism Reader: Landmarks in Literary Ecology*. Athens: U of Georgia P, 1996.

Hagan, William T. *Taking Indian Lands: The Cherokee (Jerome) Commission, 1889-1893*. Norman: U of Oklahoma P, 2003.

Hine, Robert V. and John Mack Faragher. *The American West: New Interpretive History*. NewHaven, CT: Yale UP, 2000.

Hobson, Geary. "Introduction: Remembering the Earth." Ed. Geary Hobson. *The Remembered Earth: An Anthology of Contemporary Native American Literature.* Albuquerque: U of New Mexico P, 1981. 1-11.

Hogan, Linda. *Dwellings: A Spiritual History of the Living World.* New York: W.W. Norton, 1995.

Holm, Tom. "Indian Lobbyists: Cherokee Opposition to the Allotment of Tribal Lands." *American Indian Quarterly* 5.2 (May 1979): 115-34.

Hoxie, Frederick E., ed. *Talking Back to Civilization: Indian Voices from the Progressive Era*. Boston: Bedford/St. Martins, 2001.

Jones, Robert Edmond. *The Dramatic Imagination: Reflections and Speculations on the Art of the Theatre*. New York: Theatre Arts Books, 1941.

Jones, Ruthe Blalock, Marie Depriest, and Cynthia Fowler. "Oklahoma: A View of the Center." *Studies in American Indian Literatures* 19.3 (Fall 2007): 1-44.

Justice, Daniel Heath. *Our Fire Survives the Storm: A Cherokee Literary History.* Minneapolis, U of Minnesota P, 2006.

Langner, Lawrence. *The Magic Curtain*. New York: E. P. Dutton, 1951.

Limerick, Patricia Nelson. *The Legacy of Conquest: The Unbroken Past of the American West.* New York: W. W. Norton, 1987.

—. *Something in the Soil: Legacies and Reckonings in the New West.* New York: W. W. Norton, 2000.

Maddox, Lucy. *Citizen Indians: Native American Intellectuals, Race, and Reform.* Ithaca, NY: Cornell UP, 2005.

May, Theresa. "Beyond Bambi: Toward a Dangerous Ecocriticism in Theatre Studies." *Theatre Topics* 17.2 (September 2007): 95-110.

Meacham, A. B. "Our Red Brethren—Extermination or Civilization—Which Shall it Be?" *The Council Fire* III (January 1, 1880): 1-4.

Mohawk, John. "Epilogue: Looking for Columbus: Thoughts on the Past, Present and Future of Humanity." Ed. M. Annette Jaimes. *The State of Native America: Genocide, Colonization, and Resistance.* Boston, MA: South End P, 1992. 439-44.

Otis, D. S. *The Dawes Act and the Allotment of Indian Lands*. Ed. Francis Paul Prucha. Norman: U of Oklahoma P, 1973.

Out of Dust by Lynn Riggs Program. Westport Country Playhouse, Westport, CT. August 8-13, 1949.

Pendleton, Ralph, ed. *The Theatre of Robert Edmond Jones*. Middletown, CT: Wesleyan UP, 1958.

Phillips, Sarah T. "Environmental History." Ed. Eric Foner and Lisa McGirr. *American History Now*. Philadelphia, PA: Temple UP, 2011. 285-313.

Prats, Armando José. *Invisible Natives: Myth and Identity in the American Western*. Ithaca, NY: Cornell UP, 2002.

Riggs, Lynn. *Big Lake*. New York: Samuel French, 1927.

—. *The Cherokee Night and Other Plays.* Foreword by Jace Weaver. Norman: U of Oklahoma P, 2003.

—. "A Credo for the Tributary Theatre." *Theatre Arts* xxv. 2 (February 1941): 167.

—. *Green Grow the Lilacs*. New York: Samuel French, 1931.

—. *Out of Dust* in *The Cherokee Night and Other Plays*, 212-339.

—. *Roadside: A Comedy*. New York: Samuel French, 1930.

—. *Russet Mantle and The Cherokee Night: Two Plays*. New York: Samuel French, 1936.

—. *Sump'n Like Wings and A Lantern to See By: Two Oklahoma Plays.* New York: Samuel French, 1928.

Slotkin, Richard. *The Fatal Environment: The Myth of the Frontier in the Age of Industrialization, 1800-1890*. New York: Atheneum, 1985.

—. *Gunfighter Nation: The Myth of the Frontier in Twentieth-Century America.* New York: Atheneum, 1992.

—. *Regeneration through Violence: The Mythology of the American Frontier: 1600-1860.* Middletown, CT: Wesleyan UP, 1973.

Somerset-Ward, Richard. *An American Theatre: The Story of the Westport Country Playhouse 1931-2005*. New Haven, CT: Yale UP, 2005.

Stanlake, Christy. *Native American Drama: A Critical Perspective*. Cambridge, UK: Cambridge UP, 2009.

Stremlau, Rose. *Sustaining the Cherokee Family: Kinship and the Allotment of an Indigenous Nation*. Chapel Hill: U of North Carolina P, 2011.

—. "'To Domesticate and Civilize Wild Indians': Allotment and the Campaign to Reform Indian Families, 1875-1887." *Journal of Family History* 30.3 (July 2005): 265-86.

Washburn, Wilcomb E. *The Assault on Indian Tribalism: The General Allotment Law (Dawes Act) of 1887*. Philadelphia, PA: Lippincott, 1975.

Weaver, Jace. "Ethnic Cleansing, Homestyle." *Wicazo Sa Review* 10.1 (Spring 1994): 27-39.

—. Foreword. *The Cherokee Night and Other Plays*. By Lynn Riggs. Norman: U of Oklahoma P, 2003. ix-xv.

—. "More Light Than Heat: The Current State of Native American Studies." *American Indian Quarterly* 31.2 (Spring 2007): 233-55.

—. *Notes from a Miner's Canary: Essays on the State of Native America.* Albuquerque: U of New Mexico P, 2010.

—. *Other Words: American Indian Literature, Law, and Culture*. Norman: U of Oklahoma P, 2001.

—. *That the People Might Live: Native American Literatures and Native American Community*. New York: Oxford UP, 1997.

Weaver, Jace, ed. *Defending Mother Earth: Native American Perspectives on Environmental Justice.* Maryknoll, NY: Orbis Books, 1996.

Wilmeth, Don B. "Noble or Ruthless Savage? The American Indian on Stage and in the Drama." *Journal of American Drama and Theatre* 1.1 (Spring 1989): 39-78.

Wilson, Eloise. *Lynn Riggs: Oklahoma Dramatist*. Ph.D. diss., University of Pennsylvania, 1957. Ann Arbor: UMI, 1957. 0020808.

Womack, Craig S. *Red on Red: Native American Literary Separatism.* Minneapolis: U of Minnesota P, 1999.

Worster, Donald. *Dust Bowl: The Southern Plains in the 1930s*. 1979. New York: Oxford UP, 2004.

"alterNature" in Drew Hayden Taylor's *The Berlin Blues*

Construction and De(con)struction of Contested Spaces

Maryann HENCK

Leuphana Universität Lüneburg

Prologue: Drew Hayden Taylor's Serious Silly Business

> If people did not sometimes do silly things,
> nothing intelligent would ever get done.
>
> Ludwig Wittgenstein (50e)[1]

These words may be pure Wittgenstein, but their essence is pure Taylor. As a contemporary Indigenous storyteller and Anishinaabe author, Drew Hayden Taylor is not only a man with a mission but also a medium with a message. In his trademark self-deprecating manner, he contends that his comedies have "no socially redeeming qualities" and views them as opportunities to "put some very strange people together in one place and just let them be silly together" (interview 28 August 2009). The discerning interviewer, however, knows better than to take such off-the-cuff remarks at face value. Even though the literary creation on the page and its subsequent performance on the stage might initially appear to be a matter of farcical fun, they do have a more serious side.

Assessing Taylor's serious silly business consequently calls for a tailor-made and not a prêt-à-porter approach—one that addresses crucial issues on the postcolonial agenda such as indigeneity, the sacred, and the environment as well as explores them critically from an interdisciplinary perspective. Since the subject matter of *The Berlin Blues* (2007) can be linked to discourses of Otherness in performance studies, tourism, and ecocriticism, I will examine Taylor's comedy using a multifaceted approach

1 The original quote in German appears as follows: "Wenn die Menschen nicht manchmal Dummheiten machten, geschähe überhaupt nichts Gescheites" (Wittgenstein 50).

that combines ethnographer Dwight Conquergood's framework, "Moral Mapping of Performative Stances Towards the Other," with contemporary tourism theory and practice as well as postcolonial, ecocritical approaches.

These approaches to literature tend to focus on works of a more serious nature.[2] However, all is not as serious on the postcolonial front as one might be led to believe. In their introduction to *Cheeky Fictions*, Susanne Reichl and Mark Stein suggest that "laughter is not considered a cultural attribute of a specific postcolonial space or a marker of an inherent Otherness; laughter is rather considered a device which is self-consciously employed and strategically positioned in textual constructions" (2-3). In fact, Reichl and Stein's notion of laughter as a device echoes Canadian author and environmental activist Margaret Atwood's view on Native humor as "a subversive weapon [...] for people who find themselves in a fairly tight spot without other, more physical, weapons" (244). It "get[s] the knife in, not by whacking you over the head with [...] moral righteousness, but by being funny" (244).

In *The Berlin Blues*, Taylor lets humor and laughter do the talking— not the preaching—by presenting the zealous endeavors of two overly efficient German developers from GREAT (German Recreational Entertainment, Arts & Technology) who wish to bring "civilization" and "prosperity" (91) to the economically depressed reserve of Otter Lake. In this Faustian farce of sorts, Taylor's recipe for whipping up some serious silly business could read like this: Take the following characters—Birgit, one perfectionist German entrepreneur, a wannabe with a taste for the exotic; Reinhart, one aesthetically inclined German architect, a wannabe-turns-shouldabeen;[3] Angie, one "ethical martyr" (9) and civilly disobedient alterNative activist;

---

2 For instance, in *Postcolonial Ecocriticism*, Graham Huggan and Helen Tiffin offer readings of Maori writer Patricia Grace's *Potiki* (1986) and Antiguan author Jamaica Kincaid's *A Small Place* (1988) (69-78). Additionally, Theresa J. May takes an ecocritical look at plays such as Arthur Miller's *Death of a Salesman* (1948) and Robert Schenkkan's *The Kentucky Cycle* (1993) ("Greening" 90-94).

3 In Taylor's novel, *Motorcycles & Sweetgrass* (2010), one minor Native character contrasts wannabes—"people who were [...] fascinated with Native culture and *wanted to be* Native"—with shouldabeens—"those who were unfortunate enough not to be Native but who *should have been* Native" (67 [emphasis in original]). In Act One, both Birgit and Reinhart reveal their wannabe tendencies and dilettantish fascination with all things Native by presenting the Ojibways with their concept for a profitable business venture: a Native theme park called OjibwayWorld. However, Reinhart—unlike Birgit—experiences a change of heart in Act Two and consequently rejects OjibwayWorld as well as Birgit's culturally exploitative business practices. He is "not so quick to abandon the community" and believes "there is still something to be done there" (Taylor *Berlin* 91). No longer a misguided wannabe, Reinhart ultimately shows a sincere appreciation and understanding of Ojibway culture. He gets accepted into the Ojibway fold, is allowed to stay on the reserve, and even marries Andrew and Angie (94-95). Thus, he qualifies as a potential shouldabeen.

Andrew, one harmony-obsessed police officer on the reserve; Donalda, one financially conscious, yet financially constrained head of economic development; Trailer, one unemployed Native musician, a good-for-nothing who might be a good-for-something; and approximately 400,000 pounds of imported, unionized buffalo. Then pour Birgit, a contemporary Mephistopheles, and her business partner, Reinhart, into the Ojibway[4] reserve of Otter Lake. Blend in their not-so-modest proposal for "Ojibway-tastic" OjibwayWorld (22), a C$ 164,000 Native theme park and misrepresentative cultural commodity. Add four initially skeptical Ojibways: Angie, Andrew, Donalda, and Trailer. Mix them thoroughly until they agree to sell their souls to the Germans and naively enter into a Faustian treaty. Stir things up by allowing a rebellious Angie to resist this financially enticing, yet inappropriate offer. Do not forget to go easy on the buffalo seasoning. Voilà! Now you have the one-hundred-sixty-four-thousand-dollar question that Andrew poses: "So you're saying it's either OjibwayWorld or death?" (38). Does a culture have to sell out in order not to die out?

This very serious question is tried and tested in Taylor's zany comedy. He serves up the answers in the form of over-the-top stereotypes and tourism gone wild, successfully using levity to reveal the true gravity of the situations presented. In this essay, I will examine how the commodification and ensuing exploitation of culturally as well as ecologically contested spaces lead to the construction of a hybridized "alterNature," a space ultimately destined for de(con)struction to hopefully pave the way for a more "alterNative" space—one in which culture and nature would be respected and preserved.

Act One: The Imperialists' New Clothes—Red by Appropriation, "Green by Design"?[5]

In the age of globalization, tourism, considered the world's fastest growing industry, is often depicted in diametrically opposed categories of good and evil. On the one hand, eco-, community-based, and especially pro-poor tourism have been largely praised as beneficent saviors that can create agency for the inhabitants of economically depressed regions. These approaches to tourism can benefit a wide range of stakeholders, for

4 The terms "Anishinaabe" and "Ojibway" will be used interchangeably in this paper.

5 The heading Red by Appropriation, "Green by Design" was inspired by Drew Hayden Taylor and Daniel Esty, respectively. In his essay, "Culture by Association," Taylor broaches the controversial topic of certain non-Natives who self-identify as Native by dint of marriage. He cautiously refers to this practice as "spousal cultural appropriation" and subsequently mentions the more pejorative term: "Indian by ejaculation" (117). Similarly, Birgit and Reinhart could be considered red by appropriation since they view themselves as Native on account of their "knowledge" of Native culture. In the Yale Environmental Performance Index of 2010, report author Daniel Esty praised Germany for "making its economy 'green by design'" and for its 13th-place ranking on this index ("Renewables").

instance by offering long-term employment opportunities to unskilled workers and lower-income women (Neto 212, 220). Additionally, they could contribute to the preservation of communities' natural capital and cultural heritage—often the only "assets that some of the poor have, even if they have no financial resources" (Ashley *et al.* 2). In specific, the pro-poor approach aims to "generate net benefits for the poor" which "may be economic, [...] social, environmental, or cultural" (2).[6]

On the other hand, tourism has been decried as an evil hegemonic force that inflicts rapid industrial progress, blatant commercial opportunism as well as fabricated, tourist-oriented cultural performances upon culture and the environment. In *The Golden Hordes*, John Ash and Louis Turner warned in the mid-1970s about the dangers of tourism as a form of cultural imperialism, equating "tourists with barbarian tribes" since "both involve the mass migration of people who collide with cultures far removed from their own" (11). In the 21st century, even more acceptable forms of tourism such as ecotourism, which has been lauded by proponents for its cultural and environmental sensitivity, have come under fire. As Martin Mowforth and Ian Munt have noted, opponents criticize ecotourism as "a new form of ecological imperialism in which western cultural values override local cultural values and thereby oppose the principles of sustainability which ecotourism claims to support" (107).[7]

Instead of merely pitting good and evil notions of tourism against each other, it would be useful to enter the concept of the tourist gaze into the equation.[8] In *The Tourist Gaze 3.0*, John Urry and Jonas Larsen explain: "There is no single tourist gaze as such. [...] There are many different ways of gazing within tourism, and tourists look at 'difference' differently" (2-3). Moreover, gazing is "a performance that orders, shapes and classifies, rather than reflects the world" (2). Consequently,

6 The pro-poor approach specifically focuses on increasing the local communities' involvement in and profit from touristic developments by creating business and employment opportunities to improve their overall financial independence and security. Some of objectives of this approach include the implementation of measures for managing potential social and environmental repercussions as well as the initiation of policy reforms for fostering partnerships between government agencies or the private sector and the locals (Neto 220).

7 Mowforth and Munt cite the following commonly used principles in tourism: cultural, social, environmental, and economic sustainability (100-104). For further information on the ecotourism debate, see Chapter 6, "Socio-environmental organizations: where shall we save next?" (148-76).

8 Initially influenced by Michel Foucault's "medical gaze," John Urry and Jonas Larsen use Foucault's concept to distinguish between "seeing" and "gazing." Whereas "seeing is what the human eye does," gazing is a "learned ability" that involves "socially constructed seeing" (1-2).

tourists cannot be regarded as passive consumers but are in effect active interpreters who have different readings of the touristic commodities they consume. In *The Postcolonial Exotic*, Graham Huggan notes that tourist gazes tend to be prescriptive since they do not merely "provide a medium for what tourists see, but also a guideline as to how they *ought* to see" (180 [emphasis in original]). The tourist gaze remains a problematic concept per se due to the imbalance of power relations between the gazers and the object of their gaze—the Other. Huggan also pinpoints a connection between the tourist gaze and exoticism, which engenders the "production, representation, and exploitation of the cultural other" (177). More specifically, Cara Aitchinson has criticized contemporary tourism studies for a lack of sufficient "critiques of the colonial legacy or gendered construction of Othering" as well as for the Othering of "tourist destinations as *sites* [...] and the people within them as *sights* for tourists" (137 [emphasis in original]). In their study of tourism in the Third World, Mowforth and Munt refer to this process as "zooification" since it involves treating the "hosts"—in this case, tribal people—as "objects to be viewed" (263). According to the World Council of Churches in 2002, this practice "forces Indigenous Peoples to become showcases and 'human museum exhibits'" (qtd. in Mowforth and Munt 263).

Figure 1. Moral Mapping of Performative Stances Toward the Other*

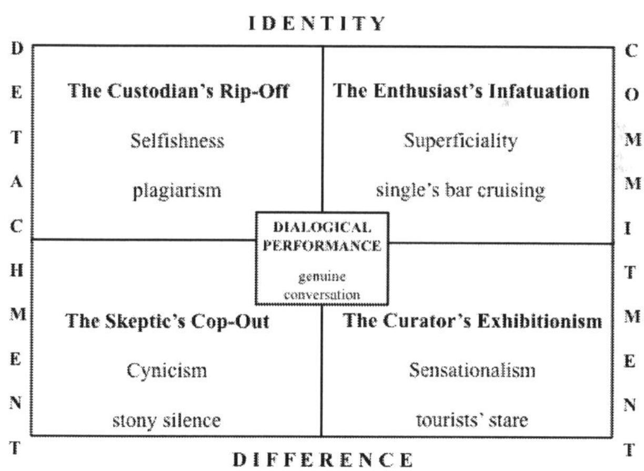

*This graphic representation is derived from Mary Douglas' method of grid/group analysis. See *Cultural Bias* (1978) and *In the Active Voice* (1982).

Source: Dwight Conquergood, "Performing as a Moral Act:[1] Ethical Dimensions of the Ethnography of Performance"; *Text and Performance Quarterly* 5.2 (1985): 1-13; *Taylor & Francis Online*; Web; 22 February 2012; 5.

Although these various discourses of tourism offer useful points of departure for addressing interactions between people and places, they need to be linked to performative as well as ecocritical approaches in the context of dramatic performances. In his essay, “Performing as a Moral Act,” Conquergood offers a useful framework entitled “Moral Mapping of Performative Stances Towards the Other” (fig. 1) for addressing the validity and problematic nature of intercultural performances between the observed community and ethnographers.

In this framework, Conquergood delineates four ethical pitfalls and offers examples of situations that create ethnographic tensions. Whereas the four quadrants depict extreme performative stances, the vertical and horizontal axes illustrate the opposing forces of identity and difference as well as detachment and commitment, respectively. In each stance, the culprit (e.g. the curator) is guilty of committing a specific sin (e.g. sensationalism), which inevitably has negative repercussions (e.g. tourists’ stare). The objective is to transcend these “binary oppositions” and gravitate towards the “dynamic center” of dialogical performance, which serves to counterbalance the four problematic stances (Conquergood 5). I will apply three of these stances to the performed intercultural interactions between “hosts”—the Ojibways—and “guests”—the German developers—in *The Berlin Blues*.

In Act One, Taylor dishes up a classic worst-practice example of The Curator’s Exhibitionism, which Conquergood also refers to as “the ‘Wild Kingdom’ approach to performance” (7). In this stance, the curators do not identify with the cultural Other but commit to and literally showcase difference by accentuating the exotic. Their sin and primary objective is to encourage sensationalism, which, in turn, activates the tourist gaze. Thus, Taylor takes the Germans’ unrivalled fascination with the “red” agenda of “Indianthusiasm” as well as the “green” agenda of environmental awareness and sustainability to set the stage for the creation of a modern-day Wild Kingdom. The developers, Birgit and Reinhart, present their theme park concept to Donalda, the head of economic development, urging her to capitalize on the Germans’ fascination with Native people. To a certain extent, this fascination can be traced back to Karl May’s popular adventure novels about Winnetou and Old Shatterhand. According to Birgit, the theme park will be a fabulous life-altering opportunity for everyone although the extent to which it will alter everyone’s lives is at first unclear. Her business partner, Reinhart, reveals the name of their brainchild as OjibwayWorld and explains: “Yes we realize you as a people prefer to be called Anishinaabe but for marketing purposes, it was decided to go with the more internationally known name of Ojibway” (Taylor *Berlin*

18). Since the Germans' motto is "Think bigger. Think huger" (18), the park will not only include amusement rides such as bumper canoes, the Medicine Ferris Wheel, and the Weesageechak Water Slide (19) but also nature-altering creations such as "a facsimile of the Rocky Mountains" and "daily caribou migrations" (20). When Donalda comments on the Germans' faulty research about the location of the Rocky Mountains and the natural habitat of caribou, Birgit counters: "While the theme park itself will be called OjibwayWorld, we will try to incorporate aspects of as many other great and proud First Nations as possible into the design of the project. After all, you are all brothers and sisters, aren't you?" (20).

Know-it-all Birgit speaks better Ojibway than the Ojibways and attempts to out-Indian the Indian; however, her knowledge has been gleaned second-hand, e.g. by reading Karl May novels, sewing her own buckskin dresses, and dancing at powwows in Germany (Taylor *Berlin* 41). In contrast to Reinhart, who has actually lived in Native communities, Birgit has only learned about and not lived Native culture. She has no scruples about collectively lumping together all Native tribes irrespective of individual differences and forcing the Ojibways to adopt a generic group identity for strategic marketing purposes. Birgit turns out to be more than just a sycophantic wannabe; she also becomes an "Otherness machine" (Aitchison 143) in her role as curator of the *site* of OjibwayWorld, her own bombastic open-air museum. Thus, she feels justified in zooifying the Ojibways as *sights* as well as expecting them to peddle touristic *sites* such as the Chief Dan George Gorge for white-water kayaking or the Crazy Horse Pony Ride (Taylor *Berlin* 79, 81), which cater to the preconceptions and escapist fantasies of non-Native theme park visitors. By romanticizing Natives as spiritual bow-and-arrow-toting noble savages and expecting them to accommodate media-cultivated stereotypes, Birgit not only reveals her unwavering commitment to her exoticized image of them, but also her utter lack of understanding of Native identity in the wake of colonization. In fact, Tzvetan Todorov rightfully criticizes the conflicting tourist gazes of that infamous colonizer, Christopher Columbus. He explains that Columbus' skewed views of Natives as both "noble savages" and "dirty dogs" arose from his "failure to recognize the Indians, and the refusal to admit them as a subject having the same rights as oneself, but different. Columbus discovered America but not the Americans" (Todorov 49). One could equally contend that Birgit discovered OjibwayWorld but not the Ojibways.

Yet, Birgit, in addition to being guilty of sensationalizing and exoticizing Native culture, also stands accused of environmental imperialism. Her telling statement, "You know what they say, you cannot make a beautiful totem pole without chopping down a tree" (Taylor *Berlin* 20), sets the tone for a devastating course of events. Birgit implies that her offer to provide the Ojibways with a veritable cash cow has a catch—one that will require sacrifices both in the name of culture and the environment. Since Indigenous identity is closely intertwined with cultural traditions and a deep connection to the land, issues involving land are often fraught with controversy—not just between Natives and non-Natives but also amongst Natives themselves.[9] In *Postcolonial Ecocriticism*, Graham Huggan and Helen Tiffin address "'nativist' and 'developmentalist' understandings of land" (69) in their analysis of Patricia Grace's *Potiki* (1986), a novel depicting conflicts between theme park developers and the Maori inhabitants of the land in question. From the perspective of nativists, "people belong to land" (Grace qtd. in Huggan and Tiffin 70) and are expected to assume stewardship for the place in which they "live, work, and play."[10] Arguing from a developmentalist viewpoint, Birgit—similar to the developers in Grace's novel—believes that "land belongs to people" (70) and can therefore be altered to accommodate their purposes and even more preferably—hers. Almost single-handedly, Birgit succeeds in infringing upon some of the most basic tenets of sustainable development and tourism, for instance: not overcrowding areas, enforcing height limits as well as valuing local culture and traditions (Plog qtd. in Burns 116).[11] For someone who clearly prides herself in being red by appropriation, Birgit's business practices are a far cry from being "green by design," thus directly contrasting with Germany's good reputation and high ranking as a green economy.[12] The message is clear: Economic profit overrides cultural and environmental issues.

In order to realize her lifelong dream and vision, Birgit must convince the Ojibways that the benefits of the theme park will outweigh any potential disadvantages. Usually, a theme park is not the first image that comes to mind as an issue of environmental inequity; more often images of toxic waste sites in poverty-stricken areas are likely to be invoked. If the choice were between using reserve land as a landfill for toxic waste

9 In *Motorcycles & Sweetgrass*, an ongoing land dispute takes center stage as the Natives on the reserve voice their varying opinions about the best way to use the newly acquired land. It could be turned into a training ground for protesters, a golf course, a casino, a gigantic movie studio, a hotel, or even a theme park catering to "Indianthusiastic" Germans (69, 123-24).

10 In this broad definition of "environment," I also am alluding to Patrick Novotny's study, *Where We Live, Work, and Play: The Environmental Justice Movement and the Struggle for a New Environmentalism* (2000).

11 These are a just few examples from Stanley Plog's PRESERVING approach (1994).

12 According to the latest Yale University Environmental Performance Index, Germany holds an 11th-place ranking and is considered a "strong performer" (*2012 EPI*).

or as the location of a new theme park, the latter would probably appear to be the lesser evil on the NIMBY (Not-In-My-Backyard) continuum. Since Birgit speaks the "language of opportunity" (Huggan and Tiffin 70), her developmentalist line of argumentation could even overturn the relatively strong nativist tendencies of the Ojibways. In other words, she lets money and power do the talking. As a result, Donalda, the economic conscience of the community, is willing to overlook potential cultural and environmental exploitation in exchange for a decade-long guarantee of financial independence as a welcome alternative to "poverty, substance abuse, and lack of a focused future" (Taylor *Berlin* 21, 36). Andrew, proponent of security and health issues, is Birgit's next target and easy prey. He gives in after hearing that the Germans intend to build a wellness center in the village, a vast improvement that could potentially combat the rising levels of diabetes and obesity in their Native community (48).

However, Trailer, "a man of little ambition and even littler means" (Taylor *Berlin* 12), is not so easily swayed. He is initially content playing in a band and living in a run-down trailer with a broken water pump. In fact, he considers his trailer a "family heirloom" even though it is in effect a *site* of "rural decay" (29), a sorry reminder of treaties signed in the aftermath of colonization. According to Malcolm Crick, much of tourism's appeal relies on a healthy portion of Otherness, but not too much because the created place should be "exotic but not alien, exciting yet not frightening" (qtd. in Huggan 178). Crick also warns, "such a space [...] requires sweeping most of social reality under the carpet" (178). In this case, it involves relocating Trailer's trailer so that the Sweatlodge Spa and Sauna can be built in its place (Taylor *Berlin* 28). Of course, Birgit offers Trailer "fair compensation" (42) by enticing him with a brand new treaty that promises a lucrative job with an impressive title, "OjibwayWorld's Chief Entertainment Development Coordinator and Artistic Producer" (43). The mere thought of money, fame, and a new water pump is music to Trailer's ears. Thus, the nativist stronghold of Otter Lake eventually begins to crumble. Angie, who has grown up off the reserve but nevertheless feels even stronger ties to the Ojibway culture than the others, finds herself in a double bind. She firmly believes that their culture should not be peddled but somehow has to "balance [...] righteous indignation with a compulsive need to eat" (9). Eventually, "reality makes hypocrites of us all" (9)—all but one, for Angie remains the nativist hardliner who speaks the "language of resistance" (Huggan and Tiffan 70), claiming that OjibwayWorld "prostitutes our culture. And we're the pimps" (Taylor *Berlin* 34).

Act Two: Travail and Travesty

If Act One depicts a ludicrous version of The Curator's Exhibitionism, Act Two presents two additional, yet equally problematic stances that both involve reframing the sacred as sacrilegious: The Custodian's Rip-Off and The Enthusiast's Infatuation. Birgit's business partner, Reinhart, shares her undeniable attraction to Indigenous cultures; however, instead of realizing a childhood dream, he wishes to leave a legacy to the Ojibways. Similar to many wannabes, custodians identify with the culture in question, yet remain detached by keeping an aesthetic distance between themselves and works of art or performances. Unlike the exotically inclined curators who revel in difference, the more objective custodians wish to preserve cultural traditions or heritage and honor the Other; however, this stance tends to be more "plunder than performance" (Conquergood 5). Reinhart's pièce-de-résistance is a laser beam dream catcher that measures exactly "forty-four metres, forty-four centimetres in diameter" (Taylor *Berlin* 53). He tells Donalda, "not only will it be good for the park, but it also pays homage to your people and their beliefs. That is why it is positioned over the eastern entrance" (54). Since she is not familiar with the symbolism of the number four (e.g. the four directions) or the legend of the dream catcher, Reinhart—the non-Native—takes the trouble to explain these aspects of Native lore to her. Donalda conveniently blames her lack of knowledge on the fact that she's "one-eighth Irish" (54). Ironically, the dream catcher—instead of filtering out bad dreams—turns into an environmentally unfriendly contraption that scorches and kills every seagull within its radius (58). Although Reinhart's intentions are good, he is nonetheless guilty of the sin of selfishness (cf. Conquergood 5). His big-bigger-Birgitesque version of a dream catcher is less of a gift and more of a "theft" (6). Reinhart has unwittingly dishonored Ojibway culture by successfully plagiarizing a sacred cultural symbol and commodifying it for the tourist gaze. Similar to Birgit, Reinhart has also failed to adhere to sustainable practices by not adequately "tak[ing] into account [...] the cultural link between humans and their natural surroundings" (Camargo *et al.* 76-77), which includes tangible (e.g. burial sites, community gatherings) as well as intangible (e.g. sacred practices, traditional knowledge) cultural dimensions.

Whereas Reinhart appropriates a popular Native artifact, Trailer, who now has a real job and a purpose in life, sets his sights on Lakota culture and decides to stage *Dances with Wolves—The Musical.* Trailer's stance can be best characterized as The Enthusiast's Infatuation, which highlights the fine line between cultural appreciation and cultural appropriation. Although Trailer himself is Native, he is not a member of the Lakota tribe. Yet, he naively identifies with the image of this tribe, which owes its popularity to the film, *Dances with Wolves*. The sin of this less "transparently immoral" stance—which Conquergood compares to the strategies of a "pick-up artist" engaging in "singles' bar cruising"—is superficiality (5-6). The initially not-so-easily-swayed Trailer now finds himself criticizing his fellow Ojibways for their lack of vision and understanding: "You don't understand the creative hell I'm in. [...] You deal with your mundane lives while I try to create art. Something that says a greater truth, a deeper statement about *our* people" (Taylor *Berlin* 78 [my emphasis]). Instead of seeing Lakota identity, Trailer's statement of a greater truth is little more than a projection of Ojibway ideals onto Lakota identity. Moreover, the chosen medium—a Vegas-style musical—is poorly suited to convey the tragic message of Manifest Destiny echoed in the songs:

> They came from the East, looking for something.
> Even they did know not what
> Strangers from afar with even stranger ways.
> They soon changed things a lot.
>
> Now they're here to stay, forever it seems
> Getting under our feet.
> It seems times have changed, for all of us
> To them we're nothing but meat. (51)

Unfortunately, Trailer, formerly "*a downtrodden bassist*," has transformed into a "*frustrated, on-the-edge theatrical producer*" (Taylor *Berlin* 56 [emphasis in original]), a business-oriented mini-Birgit of sorts, who now spends most of the time complaining about incompetent actors, expressing his contempt for unions, and obsessing about staging the perfect musical (56-58). He has been blinded by his own dilettantish commitment to trivializing the Other—a commitment that is not limited to the Lakota. In an attempt to achieve authenticity, he imports 163 unionized buffalo so that he can choreograph a majestic stampede. According to Theresa J. May, this is a prime example of how animals become a "site of the wild tamed, named, and commodified" and serve as "an example of 'eco-minstrelsy'—a reflection of human power and privilege, the Other transformed in a performance of human desire" ("Beyond Bambi" 96). Not only should Trailer have known better than to appropriate creatures sacred to the Lakota and indigenous to the Great Plains, but he also should have refrained from forcing them to perform like circus animals in a ring. Little does he realize that his sacrilegious infraction against culture and nature will not be so easily forgiven.

Photo 1. The Buffalo from 4th Line Theatre production of Drew Hayden Taylor's *The Berlin Blues* (2011). Photo credit: Wayne Eardlley, Puppet builder and operator Brad Brackenridge.

Epilogue: Mother Nature Strikes Back—Taking Back the Habitat

Even though it seems as if all is right in OjibwayWorld, all is not right in the world of the Ojibways. Angie, the one-woman blockade and protest movement, is determined to prevent the opening of this "genetically modified, bastardized, hybrid, freak show" (Taylor *Berlin* 61). Comparing the Germans' visit to Otter Lake to the Nazi invasion of Poland (61), she questions the Ojibways' sanity for allowing these imperialists to gain the upper hand: "What happened to this self-government argument that's been raging for twenty years? We've left the Canadian government for the Germans?" (72). Angie's understanding of indigeneity and criticism of neocolonialist tendencies are reminiscent of Taiaiake Alfred's (Mohawk) and Jeff Corntassel's (Cherokee) assertion that

> [t]he challenge of "being Indigenous," in psychic and cultural senses, forms the crucial question facing Indigenous peoples today in the era of contemporary colonialism—a form of post-modern imperialism in which domination is still the Settler imperative but where colonizers have designed and practise more subtle means (in contrast to the earlier forms of missionary and militaristic colonial enterprises) of accomplishing their objectives. (597-98)

Now that Angie has unmasked the Germans as the new "Settlers," she intends to hinder their "more subtle means" of acculturation by boycotting the theme park that will most certainly lead to the zooification of Ojibway culture. In an effort to protect Indigenous identity, she is determined to pull the other Ojibways over to her side so that they can push the neocolonialist Germans off Native land. Angie realizes that a sense of place can be equated with a sense of self. After convincing Donalda that the Ojibways have sold their souls, Angie encourages her to fulfill the Native women's obligation "to protect and preserve the culture" (Taylor *Berlin* 61) and reminds her that "one Ojibway woman is worth a dozen men, Native or non-Native" (64). As the comedy reaches its turning point, it is the Native women, Angie and Donalda, who actively "participate in the discourse of environmental justice by mapping the connections between social injustice, human and other bodies, and environmental exploitation" (May, "Beyond Bambi" 101). They take the lead in rebelling against the Germans who have constructed a new identity for the Ojibways by creating the new habitat of OjibwayWorld. Once again the message is clear: Habitat lost—identity lost. Habitat regained—identity regained.

Although the men need some convincing before they admit to their errors in judgment, both Andrew and Reinhart eventually have a change of heart. Only Trailer fails to see the light since he is still blinded by everything OjibwayWorld has given him, notably "four different types of water" (Taylor *Berlin* 68). As a result, he needs a four-hundred-thousand-pound, buffalo-sized argument to literally shake some sense back into him. From the very beginning, Trailer has sensed the buffalos' mistrust of him. Now he fears that they are conspiring against him—and rightly so. On opening day, their sudden, earth-shaking stampede is perceived as a "flesh tsunami" (88) and a "buffalo apocalypse" (89).

Thus, the adage "what goes up must come down" rings true in the grand finale of Taylor's play. No one, not even perfectionist Birgit, could have completely foreseen or prevented the damage caused by this "act of buffalo" (Taylor *Berlin* 92)—a natural consequence of unnatural actions. The buffalo as agents of Mother Nature reclaim Native land—the Ojibways' habitat—and humble those who have transgressed against culture and nature by demolishing Birgit's dream of OjibwayWorld, Reinhart's legacy of a dream catcher, and Trailer's travesty of a musical as well as his new trailer. Ultimately, Trailer admits, "I should have stayed in my trailer" (92). His fall from grace can be ascribed to hubris, the exaggerated pride that led him to toy with nature—a force he did not reckon with, yet a force that is clearly to be reckoned with.

It may seem as if the ending of this comedy falls short of Conquergood's objective of achieving a dialogical performance, which involves a genuine conversation with other cultures to "bring together different voices, world views, value systems, and beliefs" (9). He situates this stance at the center of the moral map in a space "between competing ideologies" and expects good performative ethnographers to "continuously play the oppositions between Identity and Difference" (9). In Birgit's case, this would be unlikely since her unrelenting curator stance precludes a dialogical performance. She has written off the Ojibways as savages who have rejected her gifts of "civilization" and "prosperity" (Taylor *Berlin* 90). In fact, she has already set her sights on the next exhibit she intends to curate: The Mohawk Mini-golf and Marineworld (91). Reinhart, on the other hand, makes an effort to change his custodian stance to one that shows more commitment than detachment to Ojibway culture. He plans to stay on the reserve and create "something smaller. More respectful [...] less garish. More grass roots" (91). Not choosing to go global like Birgit, Reinhart will go glocal[13]—in other words, think globally, act locally—to create a more "alterNative" space such as a cultural center or festival in which Indigenous traditions would be respected and the environment protected. As a result, ethically implemented glocal approaches could also lead to financial benefits for the Native community as well as underpin Conquergood's notion of a genuine dialogue. In Reinhart's new concept, Natives could take on the role of culture brokers (e.g. tour guides) who influence and enhance cultural exchanges between "hosts" and "guests" without negatively affecting the identity of the "host" society (Mathieson and Wall, qtd. in Burns 100).

What originally began as a Faustian farce seems to have evolved into a cautionary comedy that delivers a warning about what might come to pass if people are not careful about what they wish for. In fact, the fictional OjibwayWorld had a real-life counterpart as Taylor reveals in the following anecdote about his tour with the Native Theatre School:

> One of the performance stops for the theatre school was the small, quiet community of Rama, beautifully nestled on the shores of Lake Couchiching. While there, the troupe was offered the opportunity of staying at the brand new theme park the village was experimenting with. It was called, and I kid you not, Ojibway World. Picture it in your mind. This representation of an authentic Ojibway community from centuries ago contained real authentic Plains Cree teepees, Iroquois corn crushers, and all sorts of wildly interpretive buckskin outfits of no definable culture. I spent three nights sleeping in a teepee by the lake. If not accurate, it was at least fun. ("Rendez-vous with Rama" 14)

---

13 Sociologist Roland Robertson is often credited for popularizing the term "glocalization." In his essay, "Glocalization: Time-Space and Homogeneity-Heterogeneity," he discusses various perceptions of globalization as well as introduces the notion of glocalization. Robertson comments: "According to *The Oxford Dictionary of New Words* (1991: 134), the term 'glocal' and the process noun 'glocalization' are formed by 'telescoping *global* and *local* to make a blend.' Also according to the *Dictionary* that idea has been 'modelled on Japanese *dochakuka* (deriving from *dochaku* "living on one's own land"), originally the agricultural principle of adapting one's farming techniques to local conditions, but also adapted in Japanese business for *global localization*, a global outlook adapted to local conditions' (emphasis in original)" (28).

The author believes that reality tells the best stories, and theater provides the best forum for (re)telling these stories. In her article, "Greening the Theater," May regards theater as "a site of counter-discourse, resistance, and reimagining" (86), arguing that "ecocriticism must reach beyond dramatic texts to cope with performativity, with theatre's inherent polyphony, its *mise en scène*, and the dynamic meaning-making process of performance" (100). Since the proof is in the performance, I would like to conclude by examining the most recent production of *The Berlin Blues* at 4th Line Theatre in Milbrook, Ontario. Now it's time for a quick curtain call.

Curtain Call: How "Green" is the "*Blues*"?

Whether Taylor set out to create an ecodrama when he penned *The Berlin Blues* is debatable. In a recent interview, he commented: "It wasn't quite a priority when I was writing it, though it has become an issue with the outline for *The Berlin Blues* movie. There is a stronger environmental message developing" (30 January 2013). The question arises whether there are certain eco-mandates, perhaps "eco-mandments," that should be followed for a play to be classified as an ecodrama.[14] In May's view,

> ecodrama encompasses not only works that take environmental issues as their topic, hoping to raise consciousness or press for change, but also work that explores the beingness of the natural world in such a way that when we leave the theater, things around us are more alive, we listen better, and we have a better sense of our own ecological identity. ("Greening" 95)

Taylor's comedy undeniably follows the first eco-mandate, i.e., an ecodrama should address environmental issues. However, according to May's definition, one environmental issue does not make an ecodrama. Consequently, she sets forth two additional eco-mandates. The second one refers to the necessity of staging these issues to support the "polymorphously activist tendency" of third-wave ecocriticism, e.g. using literary works as a means of social transformation (Slovic 7). The final eco-mandate's objective involves encouraging the audience to reconsider and hopefully readjust their relationship to the environment. These aspects can be best explored when dramatic text meets theatrical production and nature meets culture—"the site of ecocriticism: the critical application of an ecological perspective to cultural representation" (May "Beyond Bambi" 95).

There is perhaps no better place to explore these eco-mandates than at 4th Line Theatre, a farm that playwright, actor, director, and artistic

14 For further ideas about additional eco-mandates, see May's "Some Green Questions to Ask a Play" ("Beyond Bambi" 105).

director, Robert Winslow, inherited in 1990. Wishing to give up neither this family possession nor his passion for theater, he founded 4th Line, which celebrated its 20th anniversary in June 2011 with the opening of *The Berlin Blues* (Ouzounian E6).[15] Winslow's "green-by-design" outdoor theater also follows a specific mandate: "To preserve and promote our Canadian cultural heritage through the development and presentation of regionally based, environmentally staged, historical dramas" (*4th Line*). Although Taylor's comedy does not fall into the category of historical dramas, it complies with 4th Line's mandate as a play dealing with regional issues. According to director Kim Blackwell:

> Our community right now is facing a lot of questions like those posed in the play. Sure, if we expand, it can be a great thing economically, but can you bring 2,000 more people into a town and think it will remain the same? And we're also debating the whole issue of wind turbines. They're a great thing for some folks, but not if you live right next door to one. (Ouzounian E6)

Moreover, 4th Line's mission is "to present this work in a politically responsible and culturally sensitive manner" as well as "to encourage innovation and experimentation resulting in productions that entertain and educate" (*4th Line*). These objectives reflect the second eco-mandate's call for activism and social change. The keynote in 4th Line's mission statement focuses on the notions of political responsibility and cultural sensitivity, which also correspond to Conquergood's dialogical performance as "the means as much as the end of honest intercultural understanding" (10) as well as contrast with The Skeptic's Cop-Out—"the most morally reprehensible corner of the map" (8). Conquergood argues that this is less of a "performative stance" and more of "bail-out" due to the skeptic's refusal to confront "the ethical tensions and moral ambiguities of performing culturally sensitive materials," e.g. by stating "I am neither black nor female: I will not perform from *The Color Purple*" (8). Artistic director Winslow, a non-Native, demonstrates a commendable willingness to take a necessary "empathic leap" (Conquergood 10); by staging Taylor's multi-vocal comedy, he not only gives characters an opportunity to voice their varying worldviews but also allows space to speak. May praises theater artists for leaving buildings that reflect "economic hierarchies" in their theatrical architecture (e.g. Broadway) and for taking their performances to specific sites such as the streets, the

---

15 In An Kosurko's review of the performance, the following cast members are mentioned: Cynthia Ashperger (Birgit), Herbie Barnes (Trailer), Clifford Cardinal (Andrew), Cheri Maracle (Donalda), Patti Shaughnessy (Angie), and Robert Winslow (Reinhart). For a preview of the 4th Line performance, see the following online news clips on YouTube: Steve Guthrie's "Berlin Blues" and "4th Line Theatre.mov."

countryside, or even the landfill ("Greening" 96).[16] OjibwayWorld, the embodiment of "alterNature" on the page, becomes "a glorious monster built up in the imagination of the audience through the story and the characters" on Blackwell's open-air stage (Kosurko). Winslow uses the space on the farm intuitively and has "staged plays in the forest, in the hayfield, in the barn—wherever it seemed right to do it" (Ouzounian E6). In an interview with GreenUpdate, he additionally notes, "[i]n a lot of our plays, landscape has been a big part: nature, the appreciation of nature, and home" (Fraser). In Taylor's view, the outdoor production of *The Berlin Blues* also "added a bit of originality to the story" (interview 30 January 2013). He explains that 4th Line "is a farm surrounded by other farms. There is a barn that is the backdrop to most of the shows and just off to the left is another farm on a hill. Frequently you could see horses grazing on that hill. Birds and wind overhead. Lots of sun and fresh air" (interview 15 February 2013). This non-traditional staging dissolves boundaries between nature and culture, allowing the performers, the audience, and the land—which now "emerges as a player"—to interact (May "Greening" 96).[17]

Although ubiquitous Mother Nature remains visible throughout the performance, she has been largely silenced by the invasive new Big Settler—OjibwayWorld. Justin Hiscox's soundscape underscores various scenes "with a little German electronica" or bass-player "Trailer's riff to the buffalo stampede" (Kosurko). Fortunately, Blackwell does not fall into Trailer's trap of the The Curator's Exhibitionism by importing live buffalo; instead, Brad Brackenridge was commissioned to design, construct, and operate the larger-than-life buffalo puppet that acts as Mother Nature's proxy for performing the dirty work of "alterNature's" de(con)struction. Blackwell's production ends with an enraged and raging buffalo that reunites the Ojibways with their reserve, reminding the audience that people belong to land. In other words, who we are is also where we are. Consequently, such a change in perspective could fulfill the final eco-mandate that challenges us to not only "investigate the long-standing humanist question 'who are we?' but also the urgent ecological question 'where are we?'" (May "Greening" 100).

---

16 May provides examples of site-specific theater such as the Theatre in the Wild in Seattle ("Greening" 96).

17 For instance, in Robert Winslow and Dale Hamilton's play, *The Last Green Hill* (2008), Winslow regards the hill as "one of the main characters" (Fraser). In this futuristic play, the world is divided into two factions: "people living off the land" and "people who have completely embraced technology" (Fraser). Whereas the former group is interested in preserving the land, the latter group intends to develop it. Consequently, a war ensues. In Winslow's opinion, "the play is largely about the importance of preserving green space and farmland" and therefore reflects the pressure many communities in Ontario are currently facing from developers (Fraser).

In conclusion, Taylor's play provides both food for thought and food for action since it succeeds in showing as well as potentially creating change in touristic practices, treatment of the environment, or perceptions of the Other. The closing scene of the comedy presents a humorous albeit cautionary message: "*All is silent for a second then a head or perhaps several heads can be seen peeking in a large window. They scan the room, look at each other and nod. They move towards the door that TRAILER exited, indicating they are following him*" (Taylor *Berlin* 96 [emphasis in original]). This image serves as a stealthy reminder that Big Mother or maybe even Big Buffalo is watching and forces us to seriously reconsider our relationship to the environment. Let us hope that Taylor does not stop doing silly things so that something intelligent continues to get done. Perhaps George Orwell sums it up best: "Every joke is a tiny revolution" (325).[18]

Works Cited

4th Line Theatre, n.d. Web. 30 January 2013.

"4th Line Theatre.mov." Online news clip. NorthumberlandNews. *YouTube*. 30 June 2011. Web. 30 January 2013.

Aitchison, Cara. "Theorizing Other discourses of tourism, gender and culture: Can the subaltern speak (in tourism)?" *Tourist Studies* 1.2 (2001): 133-47. *SAGE Journals*. Web. 21 March 2012.

Alfred, Taiaiake and Jeff Corntassel. "Being Indigenous: Resurgences against Contemporary Colonialism." *Government and Opposition* 40.4 (2005): 597-614. *Wiley Online Library*. Web. 28 August 2012.

Ashley, Caroline, Dilys Roe, and Howard Goodwin. *Pro-Poor Tourism Strategies: Making Tourism Work for the Poor: A Review of Experience*. Nottingham: Russell Press, 2001.

Atwood, Margaret. "A Double-Bladed Knife: Subversive Laughter in Two Stories by Thomas King." *Canadian Literature* 124.125 (1990): 243-50. *Canadian Literature*. Web. 23 March 2012.

Burns, Peter M. *An Introduction to Tourism and Anthropology*. London: Routledge, 1999.

Camargo, Blanca, Katy Lane, and Tazim Jamal. "Environmental Justice and Sustainable Tourism: The Missing Cultural Link." *The George Wright Society Forum* 24.3 (2007): 70-80. *George Wright Society*. Web. 28 August 2012.

Conquergood, Dwight. "Performing as a Moral Act:[1] Ethical Dimensions of the Ethnography of Performance." *Text and Performance Quarterly* 5.2 (1985): 1-13. *Taylor & Francis Online*. Web. 22 February 2012.

Fraser, Donald. "GreenUpdate with Robert Winslow at 4th Line Theatre." Online news clip. *YouTube*. 3 September 2009. Web. 30 January 2013.

[18] Specials thanks to Drew Hayden Taylor for always having answers to my countless questions and also to 4th Line Theatre, especially to Mark Stainback for granting me permission to use the buffalo photo.

Guthrie, Steve. "Berlin Blues." Online news clip. chextvDOTcom. *YouTube*. 30 August 2012. Web. 30 January 2013.

Huggan, Graham. *The Postcolonial Exotic: Marketing the Margins*. London: Routledge, 2001.

—, and Helen Tiffin. *Postcolonial Ecocriticism: Literature, Animals, Environment*. London: Routledge, 2010.

Kosurko, An. "4th Line opens with Ojibway World theme park show." Rev. of *The Berlin Blues*, by Drew Hayden Taylor. 4th Line Theatre, Milbrook, Canada. *The Peterborough Examiner* 1 July 2011. Web. 28 August 2012.

May, Theresa J. "Beyond Bambi: Toward a Dangerous Ecocriticism in Theatre Studies." *Theatre Topics* 17.2 (2007): 95-110. *Project Muse*. Web. 28 August 2012.

—. "Greening the Theater: Taking Ecocriticism from Page to Stage." *Journal of Interdisciplinary Studies* 7.1 (2005): 84-103. *JSTOR*. Web. 15 February 2012.

Mowforth, Martin, and Ian Munt. *Tourism and Sustainability: Development, Globalisation and New Tourism in the Third World*, 3rd ed. Abingdon, Oxon: Routledge, 2009.

Neto, Frederico. "A New Approach to Sustainable Tourism Development: Moving Beyond Environmental Protection." *Natural Resources Forum* 27. 3 (2003): 212-22.

Orwell, George. "Funny, But Not Vulgar." Ed. Sonia Orwell and Ian Angus. *The Collected Essays, Journalism and Letters of George Orwell. Vol. III: As I Please 1943-1945.* 1970. Harmondsworth: Penguin in association with Secker & Warburg, 1971. 324-329.

Ouzounian, Richard. "4th Line: Farming homegrown theatre." Rev. of *The Berlin Blues*, by Drew Hayden Taylor. 4th Line Theatre, Milbrook, Canada. *Toronto Star* 30 June 2011: E6.

Reichl, Susanne and Mark Stein. "Introduction." Ed. Susanne Reichl and Mark Stein. *Cheeky Fictions: Laughter and the Postcolonial*. Amsterdam: Rodopi, 2005. 1-23. Print.

"Renewables to Energize German Economy." *EXPORT.BY*. 16 Apr. 2010. Web. 23 March 2012.

Robertston, Roland. "Glocalization: Time-Space and Homogeneity-Hetereogeneity." Ed. Mike Featherstone, Scott Lash, and Roland Roberston. *Global Modernities*. 1995. London: Sage Publications, 1997. 25-44.

Taylor, Drew Hayden. Personal interview. 28 August 2009. E-mail.

—. Personal interview. 30 January 2013. E-mail.

—. Personal interview. 15 February 2013. E-mail.

—. "Culture by Association." *Furious Observations of a Blue-Eyed Ojibway: Funny, You Don't Look Like One ~~Two~~ Three*. Penticton, BC: Theytus Books, 2002. 117-119.

—. *Motorcycles & Sweetgrass*. Toronto: Alfred A. Knopf Canada, 2010.

—. "Rendez-vous with Rama." *Further Adventures of a Blue-Eyed Ojibway: Funny, You Don't Look Like One Two*. Penticton, BC: Theytus Books, 1999. 14-18.

—. *The Berlin Blues*. Vancouver, BC: Talonbooks, 2007.

Todorov, Tzvetan. *The Conquest of America*. Trans. Richard Howard. Norman: U of Oklahoma P, 1999.

Turner, Louis and John Ash. *The Golden Hordes: International Tourism and the Pleasure Periphery*. London: Constable, 1975.

Slovic, Scott. "The Third Wave of Ecocriticism: North American Reflections on the Current Phase of the Discipline." *Ecozon@: European Journal of Literature, Culture and Environment* 1.1 (2010): 4-10. Web. 28 August 2012.

Urry, John and Jonas Larsen. *The Tourist Gaze 3.0*. London: Sage Publications, 2011.

Wittgenstein, Ludwig. *Culture and Value = Vermischte Bemerkungen*. 1980. Amend. 2nd ed. with English translation. Ed. G.H. von Wright in collaboration with Heikki Nyman. Trans. Peter Winch. Oxford: Blackwell, 1989.

Yale University Environmental Performance Index. *2012 EPI & Pilot Trend EPI. Yale University Environmental Performance Index*, n.d. Web. 30 January 2013.

The Collapse of Worlds in Laura Shamas's *Chasing Honey*

Yvette NOLAN

Laura Shamas is a writer of mixed blood, with degrees in Theatre, English/ Creative Writing, and Mythological Studies, a published author with a dozen books in poetry, plays, essays, and critical thinking. As an Indigenous artist (she is an enrolled member of the Chickasaw Nation), she recognizes the interconnectedness of the natural world and the way we live in it. In her 2007 play *Chasing Honey*, she uses the environmental disaster of Colony Collapse Disorder—the phenomenon of bees suddenly disappearing from the hive—to parallel both the collapse of the Indigenous family on Turtle Island and the loss of Native American identity. While those two events may seem dissimilar, Indigenous worldview insists on connection. Shamas's characters struggle to stay connected to their histories and heritage, not just for their own survival, but for the survival of the natural world.[1]

The play opens with a dream/vision: a woman in traditional dress speaks in O'odham to her sleeping daughter, Sandy. Sandy is a young Tohono O'odham college student; the woman in her dream is her mother, who once was a leader, had powerful dreams but has since gone "crazy" (51, 56), descending into addiction.

Sandy is the newly elected president of the college's Native American Awareness Club, which is struggling with purpose, identity, and dwindling membership. Len, a new arrival at the college, joins not out of interest, but because an advisor has told him he can get a credit for writing a paper about the Club. The son of a Cherokee beekeeper travelling from Oregon to California to Texas to Arizona to pollinate the crops, Len is trying to cobble together an education between "chasing honey."

---

1 Since there is no single appropriate term with which to refer to the Indigenous people of Turtle Island, and since I am an Aboriginal Canadian writing about Native American people, I have referred in this essay to either nation or tribe—Choctaw, Cherokee, Tohono O'odham—when I know it. When I include myself in the population about which I am writing, I use the term Indigenous, which is current usage transnationally and/or globally. I use the term Native American where the playwright uses it, or when I am referring to a primarily American reality within the context of the play.

Both Sandy and Len are without mothers. While Sandy's mother, Kai, appears in her dreams giving her guidance, in the waking world she is a meth addict, and Sandy promises her father, Andrew, that she will have no contact with her. Len's mother "just couldn't take the road lifestyle" (14), leaving Len to travel endlessly with his father Jimmy. As in Colony Collapse Disorder, "[b]ees are vanishing into thin air, especially the queens [...]. We gotta protect the queens" (14). In a further extension of the metaphor, Jimmy says, "[o]ur poor queens are going crazy, and I don't know why. When something makes the queens sick, the whole hive fails" (14). Various theories as to the cause of CCD are floated—cellphones, electromagnetic fields, mites or a virus, pesticides, global warming—but it all boils down to the same idea: modern life is destroying the lifestyle of the bees, modern life is killing the bees. The bees are unable to adapt to the new world, and so they are dying. The message is ominous: Indigenous peoples who do not adapt to the new world will also disappear. Yet Shamas does not suggest that we must adapt or die. As Len says, "[i]sn't it better to preserve tradition that we still have, rather than sell-out? When you water it down, it loses all the flavor" (8).

These two already fragile family units are threatened by forces both within and outside the borders of their country. Sandy's father is a member of the Shadow Wolves, an all-Indigenous team that patrols the Arizona/Mexico border under the auspices of Immigration, Customs and Homeland Security. The Shadow Wolves are so adept at their work that they are being shipped overseas "to teach the art of tracking to U.S. Customs and Troops at the borders of two of the 'Stans': Tajikistan and Uzbekistan [...] focusing on smugglers of chemical, biological and nuclear weapons" (21).

Andrew does not appear to have any trouble reconciling his identity as an O'odham man with the fact that he is enforcing an arbitrary border that divided the O'odham territory. There are nine O'odham communities just south of the Tohono O'odham, separated only by the United States/Mexico border. When Sandy challenges her father about why he has to "serve and serve and serve", he cites "our famous warrior hero Luis Oacpicagigua" (32), who fought the Spanish encroachment on the O'odham lands that are now divided. The reference to Oacpicagigua is ironic; the O'odham man led the Pima Revolt in 1751 against the Spanish, but the rebellion was a failure and Oacpicagigua died within four years in a Spanish prison.

Andrew claims he does what he does "because of the ancestors, like my father, and his father before him..." (32). He is a proud Indigenous man (likely Tohono O'odham, though Shamas does not specify), but his actions seem contrary to Indigenous practices. Upon discovering that Len and Jimmy are migrant beekeepers, he declares: "They're druggies! Smugglers! Always!" (29). In spite of the fact that beekeeping in Indigenous America is one of the oldest traditional practices, Andrew seems to have no knowledge of it, or perhaps he has forgotten what he knows, which, in Indian country, means forgetting who you are. He refers to an unspecified group—beekeepers? Cherokee? Nomads?—as "[t]hose people" (29), indicating his identification with the dominant culture. Members of marginalized groups are extremely sensitive to the phrase which is commonly used to separate us/them from the larger group.

Andrew's work as a Shadow Wolf has isolated him from at least some of the communities to which he belongs. He receives death threats on his (and Sandy's) home phone, and the students of the Native American Awareness Club caution Len about any designs he might have on Sandy, suggesting that Andrew is highly protective of Sandy, and so powerful that because of him, "even the tribal elders go to jail" (20).

The traditional knowledge he does retain—how to cut sign, or how to track someone by their trail marks—is now used only in the service of the dominant culture. Len jokes to Sandy that "maybe we should get your dad tracking the bees for us" (42), but in fact Andrew is no longer tracking nature in the way of game, he is only tracking humans. Sandy says:

> My dad can look at a bush or a shoeprint in mud, and tell you how far ahead the guy is. How much he weighs, how much he's carrying, if his legs work right. Or if it's even a guy at all. (42)

Although in CCD, it is the Queen bees that "go crazy, or leave, and then the rest of the hive fails" (43), and Sandy' mother, the Queen bee of her family unit, appears to have gone crazy, Andrew's complete induction into the structure of the Shadow Wolves and its service to Homeland Security looks unbalanced, if not paranoid. His language, in his speech to the Shadow Wolves and elsewhere, is larded with expressions of fear, and male violence. He opens his pep talk to the Shadow Wolves with the suggestion that the government is going to "clone" them (21). He speaks of protecting the Canadian border, "so drug runners and terrorists can't penetrate there" (21). He closes his address with "[b]ut remember the warrior's honor. Remember the warrior's duty…." But like the confused bees, Andrew seems to have lost direction, and indeed is incapable of articulating what the warrior's duty is and the speech trails off. Sandy seems to understand at some level that her father is not the stable leader he purports to be, telling Len "I'm still in my dad's home 'cause I'm afraid he'd fall apart if I got a place on my own" (26).

Like Andrew, Len's father, Jimmy, is doing his work because of his ancestors: "I'm a beekeeper's son. And my father was a beekeeper's son. You're a beekeeper's son. We're healers. Honey is medicine. It's in our blood" (14). However, whereas Andrew's reference to the ancestors feels like a rationalization for his engagement in war, Jimmy is actually carrying on a traditional practice. Len expresses concern for his father if the bees actually disappear:

> He's a Beekeeper's Son of a Beekeeper's Son. Cherokee. It's who he is. White people lie, saying that the Cherokees got the bees from the Mayans, who got them from the Spanish five hundred years ago. But that's just a white myth, because no one can remember when the Cherokees didn't have bees. So to my dad, it's who we are. He's a sacred guardian of a long tradition, the nectar of the gods, so he's a healer. Honey is medicine. Without the bees, who is he? (25)

Len himself is straddling dual identities as a modern man seeking education and a traditional beekeeper following his Cherokee heritage. Although he has reluctantly joined the Native American Awareness Club, he cannot help but articulate his values to the membership. He challenges their profile on the Club's website, which "promotes stereotypes" (6), publishing racist and misogynist jokes. He alone seems to have an awareness of the gaze: "Just wonder what this says to the rest of the world…out of context" (7). He also objects to the prayer page that is planned, suggesting that the members are "commoditizing" (8) the culture. One of the members repeats the word, and we are not sure if he is rejecting the suggestion, scoffing at Len's use of a ten-dollar word, or a combination of both.

In spite of himself, Len joins the community, committing himself, his ideas, and his labour to the rebuilding of the group. The beekeeper's son tries to rebuild the Club the same way he and his father rebuild the bee population: with honey. He commits the fruits of his labour to the Club in an effort to lure in new members. The honey that they would normally feed to their own bees, or sell to support themselves, Len donates to the Club.

The straddling of two worlds is replayed in the Native American Awareness Club. There are only three members of the group when Len arrives, and when challenged by Len about the purpose of the Club, they maintain that it exists to "promote our culture and traditions" and "to raise awareness" (4). The primary concern of the Club appears to be a dwindling membership, paralleling the disappearing bees. Sandy, as the new President, promises to "turn things around" but the members seem incapable of saving themselves. They are aware that the existence of the Club is in jeopardy unless they can recruit members. They are themselves conflicted about the importance of the Club, when Native people face so many other challenges. Mack, the vice-President of the Club, says:

> I mean, there are so many things to fight. Big picture stuff. Our people are dying from diabetes. Dying from gangs. Dying from drugs. Dying from poverty. Dying from heat. Dying from walking on the border. Maybe, maybe we should fight those things now. The things that really matter. (75)

Mack's articulation of a series of deaths also marks the death of the holistic life of Native Americans. The very purpose of the Club, to promote Native American culture and tradition, is overwhelmed by the effort to survive. The thorny question of casinos is addressed as a strictly economic one by Sandy:

> I can tell you right now that we're for gaming here. We have a new hospital because of casinos. Half the Tohono O'odham Nation got jobs because of the casinos. We all get two grand a year because of casinos. (9)

All the characters of *Chasing Honey* are grappling with the fragmenting of Indigenous life on Turtle Island without any direct interaction with the mainstream. There are no non-Native characters in the play, although their influence is felt in every life, from the disappearance of the bees to the threat of chemical warfare overseas. The Native American Awareness Club is being colonized by the college, which wants it to open up its membership to non-Native students. Len's father takes off to meet a "scientist under the radar" (65) who is willing to trade an untested serum for bees to undertake "a controversial treatment with bee stings" (78) on his wife, who suffers from multiple sclerosis.

Shamas offers no easy answers to the question of how tradition and culture can be maintained in a contemporary world. The modern world itself is not demonized. The characters of *Chasing Honey* carry cell phones, struggle with gang recruitment, with relationships. Education is valued and sought, giving truth to the Plains people's adage that "Education Can Be Our Buffalo."[2] Sandy even articulates the thought to Len: "But sometimes you have to bend a little, you know. Things change. At some point, you gotta 'cross-pollinate' or die" (26).

How do people raise awareness of Native American culture when that culture seems to be dying or disappearing? What is implicit in the existence of the Club is the belief that art and culture and language,

[2] It is impossible to say with certainty who first stated "Education is our buffalo." As with many teachings from Indian Country, this one has been passed on orally. I have seen it used as the title of the Alberta Teachers Association resource manual, posted on a wall in the Western Union office/photograph museum in Lame Deer, Montana, and referenced by students from First Nations University of Canada in the Canadian Broadcasting Corporation series *8th Fire*. The phrase is usually credited to the Plains people, which is logical, given that they are the people who depended on the buffalo for food, shelter, clothing, tools, and with the disappearance of the animal had to find another way to survive.

that knowing one's self and one's history, are how one survives. The members of the Club, in spite of their ambivalence about continuing the club under such dire circumstances, do actually hold some traditional knowledge on behalf of their community. When Sandy is struggling with her dreams, both Mack and Heather remember for her:

> HEATHER
> Dance the celkona.
>
> SANDY
> I don't know how.
>
> MACK
> It's traditional. You remember, you enter the dream world through skipping—
>
> HEATHER
> It helps you listen to your dreams. Didn't your mom ever teach you?
>
> SANDY
> She never got around to it— (64)

Although Sandy's mother "never got around" to teaching her daughter the celkona in the waking world, she appears to her in a series of dreams/ visions that will eventually guide her to her final decision in the play.

At the centre of the play is the longest scene, which takes place at Jimmy and Len's place, and amongst the bees. The bees introduce the scene, as the stage directions indicate that "WE HEAR INSECTS BUZZING" (34) Further, the physical presence of the bees is suggested by "the shadow of a set of hives" (34). Here, at Jimmy's trailer, the portable home that allows for a nomadic existence, life is for the moment more complete, healthier. Even though Len's mother is absent, she is discussed without malice, and in fact credited with Len's affinity for learning. The mating ritual of both humans and bees is lightly referenced, with Jimmy teasing his son about taking a shower and changing his shirt: "Looking' good" (34).

Sandy is welcomed into the world of the bees by their keepers, who begin to teach her about the structure of the hive, the means of communication, and the role of *apis mellifera* in the health of food production in North America. The teachings illustrate the interconnectedness of everything. Bees are team members, all working for the good of the community, each with its own purpose. Jimmy has to keep moving because the farmers of the United States need the bees: "something like twenty billion dollars worth of crops are pollinated by migrating bees each year" (38). Every one, bee or human, has a role to play in the health of the whole. This role, this traditional role, supersedes the modern one of education to which Len aspires, and the direness of the environmental threat precludes the possibility of balancing both.

If the health and movement of the bees are central to the play, the dreams/visions of Sandy are its driving force. Just as there are six bee scenes, so there are six dream/visions in the play. Before I discuss the dream/visions in *Chasing Honey*, I need to articulate the discomfort I feel as an Indigenous woman talking about the relationship between dreams and this world. For many Indigenous people, there is less division between the two worlds; our relationships to our dreams are different. I often think of a line from the Bruce Beresford film *Black Robe* (screenplay by Brian Moore) in which the Jesuit priest, frustrated by his interaction with the Algonquin people says, "What can we say to people who think that dreams are the real world?" When I first heard the line, I suddenly understood the great abyss across which we work, as Natives and non-Natives, because, of course, dreams are the real world. In Indigenous thought, the ancestors are with us, always, all times are connected to each other, dreamtime and waking time. So often, when I write about Indigenous work, I have to remember whom I am addressing, a person who may not understand the ease with which I slip into discussing the truth in all these worlds, and at the same time resist the pressure to serve as a cultural informant, a guide to the Indigenous world(view) for the non-Native traveler. Here in Canada, where I work and live, I am especially reluctant to fill this role, because I think it is incumbent on those who live on this land to learn something of the beliefs and traditions of those who were here when they arrived.

Having said that, I recognize that I am often speaking to people who do not live on Turtle Island, and it behooves me to offer a few more cultural signposts, to be generous in explaining things that often seem to me self-evident. The dream world is so integral to the Tohono O'odham people that there is a practice to connect that world to this one, the celkona that Heather suggests Sandy dance that "helps you listen to your dreams" (64). That Heather uses the verb "listen" and not "interpret" or "understand" indicates the significance of dreams to this community. Mack, who so far has only expressed his culture through the publication of racist jokes on the Club website, says, "[y]ou remember, you enter the dream world through skipping" (65). Again, the implication in the words is that the dream world is as real as this one, and one needs only to enter it. Mack's suggestion that Sandy "remember" therefore indicates his innate belief that the knowledge is already in her.

The play opens, as mentioned previously, with a dream/vision of Sandy's mother, Kai, "wearing traditional dress, and a scarf on her head" (1). She speaks the first words of the play in Tohono O'odham, telling her sleeping daughter to "[g]o ahead and get up" (1). The dream/vision-Kai propels us into the story, telling her daughter it is time to learn, and "the time is now," although Sandy has no idea what she is supposed to learn or what it is time for. Although Kai is, in the waking world, a meth addict who threatens and cajoles her daughter into giving her money for drugs, she still has within her culture the gift of prophecy, as manifest in her dress, and her traditional knowledge, her language, and the creation stories of her people.

The second dream/vision features Kai "in a long flowing evening dress. She wears a sparkling ring of flowers on her head, like a crown" (32). This time, Sandy is a more active participant, appearing in a nightgown behind her mother, repeating bits of her mother's speech. The Kai in this dream/vision enumerates the aspects of a life she once had in this house—"This was my home. My kitchen. My bedroom. My curtains. My sheets. My coasters. My baskets" (32)—and seems to be attempting to explain how she came to quit this life of home and domesticity. The scene ends with the clearest articulation of the loss in the play, with Sandy turning on a light and speaking to the now gone Kai: "Mama? I don't remember what it felt like when you would hold me, and rock me, when I was young. And I want to. I want to. Mama?" (34)

By the third dream/vision, the bees have begun to make their way into her dreams, appearing as electric gold dots on a black robed figure, accompanied by "a BUZZING sound" (56). Worried, Sandy calls her father, who has been deployed overseas, and speaks to him for the last time in this world. She makes the connection between the disappearing bees, and his impending one. It is after the fourth dream/vision, once again of her father in his black robe with "electric golden dots all over it" (61), that the members of the Native American Awareness Club remember the celkona for Sandy. In his book *The Tohono O'odham and Pimeria Alta*, Allan J. McIntyre has assembled a wealth of photographs from the Arizona Historical Society, including photographs of celkona dancers, and a description of the celkona, which comes from the O'odham word "celka," to skip. According to McIntyre, "the dance and accompanying songs represent the living embodiment and interpretation of a dream. The dreamer, usually a singer, teaches young dancers the proper steps to perform and sees to the creation of his dream vision in the form of figures" (McIntyre 91). Here the threads between this world and the dream world are woven ever more tightly.

In the fifth dream/vision, her mother sings Sandy's history to her, both her own immediate history, encompassing her mother back to her great-grandmother, and the history of her people, invoking the creation story of the Tohono O'odham people. Through her song, she connects her family to the land, to the traditions and practices, to the work of growing and harvesting food, and then further back to Brother I'itoi and the Baboquivari

Peak, the sacred site central to the Tohono O'odham cosmology. Sandy repeats the lines of her mother's song, in her speaking voice. She is learning from the dream/vision Kai, and she is charged with the directive "So you will not let our tradition die" (71). This vision marks the last time Kai will speak in either world in the play, and she does so in her native language, speaking three final lines in Tohono O'odham, and then skipping away, leaving her daughter calling, "Mama? Don't leave me yet. It's too soon. It's too soon. There's so much I haven't learned—Mama?" (71)

Immediately after the vision, and the teaching, is the shortest scene in the play: Sandy, dressed only in a shirt and shorts, walks into the bees. She has chosen the bees, and service to the bees; she wants to be stung multiple times in an effort to become immune. This scene has the feel of a ceremony. Sandy knows from Len and his father that she could die if she has enough of an allergic reaction, and that most beekeepers acquire immunity gradually, over time. Nonetheless:

> She takes a deep breath. Then….
>
> SANDY walks into a dark area. The LIGHT GOES WHITE ON HER.
> BUZZING NOISES ARE HEARD. She is shrouded in black light.
> A SCREAM. It ECHOES. BUZZING SOUND GROWS. CRESCENDOS. (71)

One after another, Sandy's worlds collapse. The college wants the Native American Awareness Club "to recruit a certain percentage of non-Indians by next year. To prove our club benefits the entire school" (74). After an effort to stave off colonization, a protest at the Provost's office that literally goes unheard by anyone in power, she disbands the Club. It is in this speech to a gatekeeper of some sort—a secretary or assistant whom we never see—that we learn that the Club was started decades earlier by her parents at another College. Even as Sandy is invoking her parents and their role in the creation of the Club, they are markedly absent. There is news of a wreck on the highway in which a woman is killed, and although Shamas does not elaborate, both Sandy and the audience suspect the casualty is Kai. Sandy's father, somewhere in "the mountains in Pakistan" (58) is not calling, out of "communication range" (81). She is, to all intents and purposes, an orphan.

In the penultimate scene, her parents appear to her in her dream/vision. The very fact that they appear together is alarming for Sandy, and confirms for the audience that both parents are no longer in this world but have passed over. For Sandy, the confirmation is that her mother is wearing the basket on her head, fulfilling the prediction that it would signal her death. Andrew is once again in the black buzzing golden-lit robe. They do not speak, and disappear when Sandy tries to approach them.

The play ends with the two young people heading north with as many bees as they can transport. Len, also alone since his father chased off after a serum that may or may not avert Colony Collapse Disorder, agrees to take Sandy with him, turning his acquiescence into an invitation so formal it sounds like a pact. As if at a wedding, the parents of the bride appear, "wearing all white" (82):

> LEN
> I hereby invite you to come help us try to save the crops and the bees, and learn how to make honey, while we still can. Before it's too late.
>
> SANDY
> I accept. (82)

Then, in an impromptu ceremony, Sandy asks for the bees to bless her journey with Len, as well as her parents' journey. The last words of the play are in Tohono O'odham, as Sandy turns to each of the four directions. The text she quotes is from a ritualized speech of the Tohono O'odham people, recorded by Ruth M. Underhill in the book *Rainhouse and Ocean—Speeches for the Papago Year*, and identified as a "Starting Home Speech" (Underhill *et al.* 128). The lines that Sandy speaks in her ceremony are the lines that immediately follow the lines Kai spoke in the teaching dream/vision, from the same Starting Home Speech. In this way, the teaching goes on, over time and times, from generation to generation.

All the characters are chasing honey, each in his or her own way. Jimmy and Len believe that honey is medicine, and that they are healers. They hope to heal the earth with the bees. Jimmy is attempting to revitalize the bees in the role of healer across culture, both of Indigenous America and Western medicine. Kai chases honey across realities, from the dream world to the waking world, pollinating her daughter with the knowledge from their ancestors. Even Andrew chases honey across the globe, training others who, like him, would protect their lands.

Whether Laura Shamas's *Chasing Honey* is received as hopeful or hopeless will probably depend on each individual audience member, and his or her own personal belief system. *Chasing Honey* is an alarum from an Indigenous playwright who fears for the health of the land in which she lives. She offers no solution to CCD or to the parallel collapse of the Indigenous family. We are, however, offered several windows of hope. The play ends with a wedding of sorts, although each of the marriages referenced in the play, from Sandy's parents to Sylvia Plath and Ted Hughes (until Len, everything Sandy knows about bees she has learned from Sylvia Plath), failed either spectacularly or quietly. The two young people who head off together create a new family unit, one that binds together tradition and contemporary living. The bees upon which so much depends still exist at the end of the play, both in the care of the beekeepers, and free in the Arizona landscape. How will these survivors—human and insect—do? As Len says, "[d]epends on who they mate with, if they can build a new hive. I'd give them a fifty-fifty chance. Ready?"(83).

Works Cited

McIntyre, Allan J. and the Arizona Historical Society. *The Tohono O'odham and Pimeria Alta.* Charleston, North Carolina: Arcadia, 2008.

Moore, Brian. *Black Robe*. Unpublished screenplay, 1991.

Shamas, Laura. *Chasing Honey*. Unpublished play, 2007.

Underhill, Ruth M., Bahr, Donald M, Pancho, *et al. Rainhouse and Ocean, Speeches for the Papago Year*. Tuscon: College of Arizona, 1979.

Eric Gansworth

Dramatizing the Ecology of Haudenosaunee Creation

Nicholle Dragone

Black Hills State University

According to the late Seneca scholar John Mohawk, the ecology of Haudenosaunee creation is unique among Indigenous traditional teachings about creation in that "many origin stories describe a physical world familiar to its audience, including islands or mountains or rivers which compose a living landscape," while the Haudenosaunee creation does not (Mohawk "Forward" vii). He continues to say that this creation account could have happened almost anywhere (Mohawk "Forward" vii-viii). Likewise, in his play *Re-Creation Story*, produced in 2008 at The Public Theater (New York City), Onondaga playwright Eric Gansworth uses the concept of undefined place/space inherent in the ecology of Haudenosaunee cultural traditions of creation. In doing so, he shares with his audience the relevance of Haudenosaunee epistemologies through dramatic re-invention. Indeed, as both the main character and the playwright, Gansworth invites us, as his audience, to join him as he moves from of a sparsely designed set of a college classroom to "other unstable physical places" (3) that become, for Gansworth, spaces that define the ecology of Haudenosaunee creation and re-creation. This essay seeks to explore the ways in which Gansworth juxtaposes staged space with the place/role of the audience, the imagined re-creations of the contemporary college campus, Haudenosaunee communities, and the "unstable [...] places" of Haudenosaunee creation, and how these juxtapositions of staged space with the "unstable physical places" that define the ecology of Gansworth's cultural landscape serve as a forum for the presentation of deep insights about the role of creation and re-creation in Gansworth's identity as a Haudenosaunee man and his understanding of his place in the world.

As inferred above, playwright Eric Gansworth is a citizen of both the Onondaga Nation and the Haudenosaunee Confederacy (a.k.a the People of the Longhouse, the Iroquois Confederacy and the Six Nations

Confederacy). Though a citizen of the Onondaga Nation, Gansworth was born and raised on the Tuscarora Nation, both of which are today member-nations of the Haudenosaunee Confederacy. According to traditional Haudenosaunee teachings, a man referred to today as the Peacemaker brought a message of peace to the warring nations of the Northeast, uniting the Mohawk, Oneida, Onondaga, Cayuga and Seneca Nations together into a confederacy. Traditional Haudenosaunee territories ranged from the Adirondack Mountains and the Champlain corridor of eastern upstate New York westward to the Niagara River in western upstate New York, northward into southern Ontario and Quebec and southward into northern Pennsylvania. In the early 1700s, after losing a war with the colonial settlers of North Carolina, the people of the Tuscarora Nation petitioned the Haudenosaunee for the right to relocate to Haudenosaunee territories. Eventually, the Tuscarora Nation was adopted as the sixth nation of the Haudenosaunee Confederacy. At the close of the American Revolution, the Seneca Nation gave the Tuscarora the lands comprising their current reservation on the Niagara escarpment, which is located near Niagara Falls, New York. At the time the Tuscarora relocated to this new homeland, several Onondaga women married to Tuscarora men moved with them. Citizenship in both the Nation and the Confederacy is inherited through the mother, so the descendants of these Onondaga women remain Onondaga citizens residing on Tuscarora Nation lands. Eric Gansworth's family is descended from one of these Onondaga women.

It is important to note, however, that while the Tuscarora Nation was adopted into the Confederacy, and while the Tuscarora people do participate in the traditional Haudenosaunee system of government, there are few people at Tuscarora who participate in the ancient Haudenosaunee spiritual traditions (a.k.a the Longhouse tradition). Eric Gansworth and his family were no exceptions to this reality of modern life at Tuscarora. Indeed, as an Onondaga resident on Tuscarora lands, Gansworth grew up, by his own admission, outside of the ecology of the Longhouse, and, therefore, outside of the ecology of Haudenosaunee creation. The events dramatized in *Re-Creation Story* explain how Gansworth stepped into the Haudenosaunee Longhouse ecology by exploring the cultural landscapes of the Haudenosaunee creation teachings, and how his journey through these landscapes, in Gansworth's own opinion, redefined his identity as a Haudenosaunee man and his understanding of his place in and relationship to the world.

As both the playwright and the character narrating this story of re-creation, Eric Gansworth chooses to juxtapose a sparsely set stage with *his* imagined re-creations of *his own* contemporary college campus; the Haudenosaunee communities with which he, as an Onondaga man growing up on the Tuscarora Nation's reservation, is relationally accountable to; and, the "unstable [...] places" of Haudenosaunee creation, more specifically Sky World and Turtle's Back. Most of the action in this play takes place on the college campus where Gansworth-the-character is employed as a Professor of English. Downstage, the stage is furnished with a podium; and, upstage, with a table, one chair, a laptop and a lamp. Five other chairs are positioned in two semi-circles between the podium and the table. This set functions as the college classroom where Gansworth delivers his creation talk, as well as his campus office. In addition, the stage is set to portray the kitchen in his niece's house on the Onondaga reservation (3, 19). However, the stage directions call for the use of dry ice throughout the play. The intention here is to juxtapose Sky World and Turtle's back—locations of creation—with Gansworth's memory of the various places on his college campus where the action is set. He describes these locations as "unstable" physical spaces, spaces of reflection and speculation.

The portrayal of characters other than Gansworth also augments the stage setting by enhancing the unstable mythical spaces and tenuous, shifting grounds of reflection and speculation where the play takes place. To do this, Gansworth not only refuses to identify these characters by name, he also keeps them shrouded in a sustained cloud of dry ice and silhouetted by backlighting. Thus, both the physical setting and the anonymous characters that Gansworth interacts with are "nebulous" and ethereal (3), as if existing either in the playwright's memory or in the mythical space of Creation.

Moreover, Gansworth-the-playwright adds yet another twist to his staged space, for throughout the play, he projects a series of slides onto a screen behind the actors. These projected images enable both the theater audience and the staged audience to visualize various elements of the creation story when Gansworth's knowledge of it and/ or his ability to function as a traditional storyteller are still too limited to fully engage with the story and his audience in any meaningful way. In fact, Gansworth-the-character tells the people who have come to hear his creation talk that he is intentionally using a Powerpoint presentation to create the ecology of this cultural landscape necessary to tell this story:

> I have a, um, a Powerpoint here, with some paintings of mine. They trace the months of the year, or moons, as our calendar designates them [...] this series shows the moon's progression through the year [...]. Our year starts in what you call February, for us, the Mid-Winter Moon. (6)

During this first telling, while Gansworth stumbles through the events of the story and an actress portraying Sky Woman pantomimes her fall from the sky and dances Turtle's Back into Turtle Island,[1] images of Sky Woman and her grandsons are projected behind the action (7-8). The slides fill in the gaps left by Gansworth's lackluster presentation and Sky Woman's halted interactions with the storyteller. Later, when Gansworth-the-character bemoans all the unsolicited advice he received after agreeing to tell the creation story, an advertisement is projected on the back-screen. This advertisement suggests Gansworth should contact Sky Woman for help: "*Onscreen, advertising font, with Call-1800-SKY-WOMAN*" (17). In later scenes, slides depicting various aspects of Haudenosaunee creation are again projected as Gansworth tells the creation story for a second and third time (46). Some of these slides present "*images from a 'graphic novel' style version of the story of Skywoman [...] Skywoman's husband dreams then awakens [...] he leaves the hole time sequenced [...] Skywoman's landing click sequenced*" (50-51).[2]

Gansworth also projects images that are not culturally specific but help him tell his story by juxtaposing staged space with the audience's space, often creating a place where dialogue about the creation can unfold. For example, towards the end of the play, an image of dandelion seeds is projected overhead while a group of backlighted, silhouetted characters engage with Gansworth in a dialogue about the symbolism of dandelion seeds. Eventually, the seeds are referred to in much the same way as the unsolicited advice Gansworth received on telling the creation story:

> ERIC: After I gave this talk the first time, the classroom disaster, an Indian man who had been in the audience came up and said—.
>
> ADULT MALE: You know I hear dandelions, they're an invasive species. They came over from Europe. And now we can't get rid of them.
>
> ERIC: That seemed about right. But no matter where they come from, those seeds, as they spill out on breezes, they
>
> ELDER ADULT MALE: keep it simple,

---

1 In the Haudenosaunee creation account, the water animals and birds realize that this woman who just fell through a hole in the sky cannot live in the water, nor can she live on Turtle's back. The water birds and animals council together in an effort to determine how they can help Sky Woman to survive in what for her is a new and alien water world. As a result of this council, Muskrat dives down to the depths of the ocean and brings back dirt for her. Sky Woman takes the dirt, places it on Turtle's back and dances clockwise on the dirt, around Turtle's back. As she does so, both the dirt on Turtle's back and Turtle itself grow in size until they become what Haudenosaunee people now call Turtle Island (North America). Today, Haudenosaunee women still shuffle-dance in honor of Sky Woman's act of creation.

2 Gansworth is an accomplished visual artist; he painted the "graphic novel" style version of Skywoman's story. Some of his more recent artwork can be viewed by exploring his website www.ericgansworth.com.

> ELDER ADULT FEMALE: stick to the basics,
>
> ADULT FEMALE: tie themselves to ceremonial life,
>
> ADULT MALE: tie themselves to everyday life. (75)

An ADOLESCENT FEMALE ends this conversation by commenting that the dandelion seeds tied themselves to everyday life after falling out of Skywoman's dress, into the universe. At that moment, the slide of the dandelion seeds fades to an image of *"the moon and the universe"* (75). This is significant because Gansworth-the-playwright opens both the play and Gansworth-the-character's first creation talk with the projection of an image of the Haudenosaunee Mid-Winter Moon, which Gansworth equates both to the beginning of the Haudenosaunee year and to the month of February:

> *Time, February 2008*
>
> *Scene 1:*
> *(Onscreen: Slide 1: "Re-Creation Story, February 2008, or in the Haudenosaunee calendar, the beginning of the year, AKA" with music cue, "Stick Dance, which plays throughout the next two slides dissolves to*
>
> *Slide 2: click at: 15 "The Mid-Winter Moon" title card, dissolve to*
>
> *Slide 3: click at: 3, an image of "The Mid-Winter Moon" as painted in Nickel Eclipse.)*[3] *These title cards and images, in order, will make up the opening of each scene. They show a visual narrative progression of months and seasons and an "eclipse."* (4)
>
> Eric: [...] Our year starts in what you call February, for us, the Mid-Winter Moon. (6)

At periodic intervals throughout the play, images of the rest of the Haudenosaunee's calendrical moons are projected behind the actors: the Sugar Moon, the Fishing Moon, the Planting Moon, the Strawberry Moon, the Blueberry Moon, the Lost Moon, the Green Corn Moon, the Fresh Moon, the Harvest Moon, the Hunting Moon, the Cold Moon, and the Very Cold Moon, (19, 22, 34, 39, 46, 47, 49, 55, 57, 59, 64, 68). The projected images of these moons outline the cyclical nature of the Haudenosaunee calendar/year by reminding the people that it is time to come together and partake in the Longhouse thanksgiving ceremonials given to the people by the Good Mind during the creation. These ceremonials correspond to the moons of the Haudenosaunee calendar:

> ERIC: [...] Before the brothers leave, for the celestial pathway back into the Skyworld, the Good Mind gives the humans the ceremonies.

---

3 Eric Gansworth is also a published poet, novelist and visual artist. The painting referred to here, "The Mid-Winter Moon" is printed in his poetry collection entitled *Nickel Eclipse.*

> ELDER ADULT FEMALE: It's with these that we remember who we are. They're ceremonies of thanksgiving. We understand that the chances of everything in this world working out as they have are such statistical improbabilities, that we are thankful for the beings who made it so. We are asked in our thankfulness, we live a good life, as a way of honoring that gift.
>
> ADULT FEMALE: Through hundreds of years of outside influence and pressures and the loss of language and culture, and the Dawes Act and forced assimilation, the ceremonies still go on. In this story, even our own forgetfulness becomes a part of the plot, so we will remember to be careful. Remember, in an oral culture, we are always one generation from disappearing altogether, and that we remember at all is something for which we should give thanks. (54-55)

The on-screen projection of these calendrical moons is a reminder of the ceremonies of thanksgiving, which, in turn, is a reminder that the circumstances ensuring life on Turtle Island could just as easily not have occurred. Therefore, the ceremonies, like the slides of the calendrical moons which bring them to memory, stabilize the shifting grounds of memory, perception and mythic time/place constituting Gansworth's staged space.

In some scenes, rather than projecting an image of one of the Haudenosaunee's calendrical moons, a picture of a full moon is projected on screen. Not only is this important because the projected images of the Haudenosaunee moons provide Gansworth and his audiences with a somewhat more solid, less shifting ground, but also because it again brings us back to Sky Woman's story, which is itself a story of re-creation. Again, a group of backlighted, silhouetted anonymous characters engage in a dialogue with Gansworth-the-character/storyteller about Sky Woman, her daughter, her twin grandsons and the moon:

> ELDER ADULT MALE: Sometimes, it's the Bad Mind who's attentive to his mother's grave, but he takes her body, beheads her, and hides her body in a cave, limiting the possibility that she might come back. (56)

In these versions of Haudenosaunee creation, the head of Sky Woman's daughter is eventually hung in the sky, and she becomes Grandmother Moon. However, as we find out later in the play, there is another version:

> ADULT MALE: Sometimes, it's her head that gets thrown up into the sky to become the Sun, and it's her bosom that becomes the moon. There's no daughter and no Elder Brother, and yet the sky is filled, day and night, with those involved in our Creation. (56-57)

In much the same way that Grandmother Moon becomes an integral part of Haudenosaunee culture and philosophy, the projecting of the creation story and other aspects of Haudenosaunee culture on screen become—for both Gansworth and his audience(s)—a place of refuge when all other mythic and memory grounds are shifting and unstable. It is Grandmother Moon, after all, who controls women's monthly cycles and their ability to procreate and/or re-create. It is also Grandmother Moon, who, in the form of the various moons listed above, reminds the people of their responsibility to join in the thanksgiving ceremonies.

These slides also work to juxtapose the staged space of Gansworth's *Re-Creation Story* with the imagined re-creations of his contemporary college campus, Haudenosaunee communities, and the "unstable [...] places" of Haudenosaunee creation. The slides also help us to understand how these juxtapositions of staged space, with both the "unstable physical places" and Gansworth's various audiences, provide deep insights about the ecology of Haudenosaunee creation—and its role in transforming Gansworth's identity as a Haudenosaunee man and in helping him to understand his relationships to both the Haudenosaunee world and the world at large.

Again, as described by Gansworth, the set is "constantly shifting", moving in and out of "reflection, speculation and other unstable physical places" (3). This allows Gansworth and his theater audience to be juxtaposed with a mythical place and time where Sky Woman and her twin grandsons interact with Gansworth as he shares their story with his audience(s) over and over again. Indeed, with each telling, Sky Woman and her twin grandsons take on more active roles, joining Gansworth on stage in a way that requires both the storyteller and his audience to unite with Sky Woman and her family in the process of creating and re-creating the world. In her first appearances on stage, Sky Woman is backlighted and silhouetted, like many of the other characters in this play:

> *ELDER ADULT FEMALE, backlighted, stands pantomiming falling and then a kind of floating, with hands spread at her waist, as if perhaps hang-gliding* [...] *ELDER ADULT FEMALE, still backlighted, dances THE WOMEN'S SHUFFLE DANCE, a social dance, not a ceremonial dance* [...] *GANSWORTH ignores what is going on behind him.* (6-7)

With each telling of her story, Gansworth's knowledge of creation and his ability to function as a traditional storyteller grow. With each telling of her story, Sky Woman's presence on stage becomes more solid, she takes on a more active role as a co-storyteller. In this way, she expands the scope of her dialogue and action outward to include not only Gansworth but also the theater audience, as she dances Turtle's Back into Turtle Island:

> ELDER ADULT FEMALE: I dance in thankfulness at my survival (*ADULT FEMALE DOES WOMEN'S SHUFFLE DANCE, a social dance not a ceremonial one...*) and each circuit I make around Turtle's back, the shell expands. (51)

At some point, she is joined on stage by her twin grandsons. By the end of the play, the three Sky World beings periodically interrupt Gansworth's dialogue to inform the audience that they are continuing their role in creation. For example, half-way through the play, the elder twin, Good Mind, tells the audience that he is making the world ready for humans by creating flowers, game animals and edible plants. Bad Mind, the younger brother, also tells people how he has contributed to the ongoing processes of creation by making carnivorous animals, poisonous plants, rapids, thorns and weeds (52). In the spirit of competition that marks their ability to keep this newly created world in balance, the twin minds eventually take command of the stage to tell their own story, each from their own perspective:

> ELDER ADULT MALE: I govern the day.
>
> ADULT MALE: I govern the night.
>
> ELDER ADULT MALE: I make things to satisfy, please and provide for the people I will eventually create.
>
> ADULT MALE: I teach those same people that they'll have to deal with adversity, and that there are benefits to learning to be tough.
>
> ELDER ADULT MALE: I invent the open admissions college.
>
> ADULT MALE: And I invent the circumstances that necessitate an open admissions college. (52-53)

The twin minds of creation so completely hijack Gansworth's play that he, as the main character, is only able to briefly interject that these brothers—Good Mind and Bad Mind—though at odds with each other "ultimately counter and complement the work of one another" (53). With each telling of creation Gansworth offers his audience(s), with each of his interactions with Sky Woman, Good Mind and Bad Mind, the relationship binding them all together grows stronger through their juxtaposition between the imagined re-creations of the contemporary college campus and the unstable places of Haudenosaunee creation. In fact, with each telling, Gansworth strengthens his relationships with them and with the creation story itself— relationships defined by relational accountability to the story, to the Sky World beings populating his story, to the people he is telling this story to, and to the culture the story grew out of.

Sky Woman and her grandsons are not the only ones who hijack Gansworth's telling and re-telling of creation. As Sky World and the shifting, unstable grounds of Gansworth's staged memories become more nebulous and ethereal in their juxtaposition with Turtle Island, the audience finds itself developing relationships with the characters and the sparsely designed set of *Re-Creation Story.* Their roles, as the theatrical production's audience, become entangled with and complicated by Gansworth-the-storyteller's relationship with the staged audience. One ADULT MALE tells Gansworth that it is the audience's job to correct him, to correct the storyteller (73). This man also reminds Gansworth-the-playwright that he is not actually up on the stage telling the creation story. Rather, a professional actor was hired to play the role of Gansworth-the-character. It is time now for the playwright to take his place in the audience and let the professionals do their jobs. ADULT MALE continues by admonishing Gansworth for being too stubborn to tell the story with those natural variations that oral tradition is resplendent with:

> Let's face it, it is common sense that a professional is always going to do a better job [...] But because you are so stubborn, you almost walked away from this one in the middle of the process, because you keep telling yourself the same story. But [Creation] is a story that invites variation and new lessons with each telling. So man, sit your ass down in the audience and let us help you where we can. Why is this so hard for you? (73)

Though annoyed with these interruptions, reminders and admonishments by the audience, Gansworth recognizes that he is not in line with Haudenosaunee cultural values and apologizes for trying to retain such tight control of the telling and re-telling of creation in the sense that he nearly refused to let anyone else contribute to the telling of the story (73). In the juxtaposition of the place of the audience with the place of the traditional storyteller; in the juxtaposition of staged space with the audience; in the juxtaposition of Sky World and the cultural landscape of memory, Eric Gansworth successfully undergoes a metamorphosis from novelist/poet/dramatist/visual artist to traditional Haudenosaunee storyteller. In addition, he and his audience have become active participants in the ongoing processes of creation and re-creation.

This play is the dramatization of the playwright's own journey moving from his familiar, well-defined ground of storytelling via novels/poems/ visual arts to the unstable, shifting grounds of playwriting and oral story telling. Gansworth's journey begins with a real-life request from one of Gansworth's colleagues to act as one of the featured speakers in a literary speaker series at Canisius College by giving a three-hour talk on Haudenosaunee creation (12-13). As Gansworth explains, he agrees out of a sense of obligation because his colleague was, in a large part, responsible for Gansworth's appointment as an English faculty member at Canisius (13-14). However, as the time for the presentation draws closer, he realizes how little he knows about Haudenosaunee creation apart from the few elements of the teachings he incorporates into his literary and visual arts. Gansworth sends a quick reply to an email reminder about the date of his presentation, asking whether the date of the event is correct. In the email, he comments that he "still [does not] know how the hell [he] is going to fill three hours with the Creation Story" (14). Moments later, he realizes that he accidentally replied to the Indian Higher Education listserv reminder rather than to a personal email. To his dismay, "all" of Western New York, specifically those Haudenosaunee people who subscribed to this listserv, are now aware of the fact that Gansworth does not know enough about his own creation story to talk about it for three hours (12-13). In *Re-creation Story's* dramatization of these real-life experiences, Gansworth-the-character later rationalizes his insecurities about his lack of knowledge and inexperience with oral storytelling:

> And let's admit it, some of you came here, expecting me to f__k up because you read that email. So maybe that's where this starts. I knew, of course, that I was doing this presentation a while ago [...] I was practicing at home, reading books and archive facsimiles for weeks, trying to get ready. Not only do I not know how to give a formal lecture, but I don't really know, easily, half this story. There's the whole extended first section that takes place in the Sky World that I hadn't even heard of until I read this version that came out last year. I kept reading and reading and thinking, who ARE these people. When Sky Woman finally showed up, I didn't even recognize her until she went through the hole. (11-12)

From this point on, Gansworth moves through the shifting landscape of his memory from the lecture hall where he presented Haudenosaunee creation, to his office, and to his niece's house while simultaneously alternating between telling and re-telling the creation story to his listening audience and talking to a variety of silhouetted and largely anonymous people about his responsibility to present the creation story in a college speaker series. As Gansworth takes his audience on this journey, the shifting landscape of his memory is juxtaposed with the unstable grounds of Sky World, the mythic space where the Haudenosaunee creation story originates. Through this juxtaposition, Gansworth and his audience are forced to interact with Sky Woman and her twin grandsons. As indicated above, we, as Gansworth's audience, are privy to brief conversations and portions of emails Gansworth-the-character receives from colleagues, family and friends giving him the unsolicited advice on traditional storytelling and, more specifically, telling the creation story:

> Keep it Simple [...]
> Stick to the Basics [...]
> It's the Creation story, and you'll be telling it on a Dark Winter Night, when you are supposed to be telling it. What could be more perfect? [...]
> Three Hours! You won't even have time to get to the creation of the earth in that time! Don't worry about it [...]
> Tie it to Ceremonial life [...]
> Tie it to every day Life [...]
> Ask X [...]
> Ask Y [...]
> Have you read A's work? [...]
> Have you read B's work? [...]

> Apologize for every error you are about to make. That's the traditional way. (16-17)

We are also privy to longer conversations that, in his eyes, confirmed his own inadequacies:

> ERIC: When did you come in?
>
> ELDER ADULT MALE: Long enough ago to know you're ready to start bitching about people asking you to do things you don't do well. Why did you agree to this anyway? It's not like storytelling is even remotely close to what you do.
>
> ELDER ADULT MALE: A crash course in monologue is not going to get you anywhere three weeks before you have to deliver it. Just because you're Indian, that does not make you an expert on all things Indian. (13)

By making us privy to these conversations, Gansworth-the-playwright gives us deep insights into the unstable ground Gansworth-the-character is stumbling across as he falls deeper into anxiety and emotional distress in his preparation for this presentation. This is important, because Haudenosaunee cultural identity is predicated on the concept of the "good mind". Good mind[4] is both a name and a philosophical concept that defines a way of life. At its heart, it is a peaceful, healthy mind—a mind free from the snares of grief, hatred, envy, jealousy, bitterness, etc. It is a mind that is able to contribute rationally to the ongoing processes of creation and re-creation.

In its earliest incarnations in Haudenosaunee traditional teachings, Good Mind is one of the names of Sky Woman's eldest grandson, "He-Who-Formed-Our-Bodies," or the creator of humankind. Haudenosaunee creation is, in many ways, Sky Woman's story. As her story unfolds, first in Sky World, the tree that provides light for Sky World is uprooted and a very pregnant Sky Woman either falls or is pushed through the hole made by its uprooting. She finds herself tumbling through a dark void to a water world below her. With the help of water animals and birds, Sky Woman is set on the back of a giant Turtle. Some of these water animals sacrifice themselves to bring back dirt, which their oral traditions tell them may be found on the ocean floor. Sky Woman places this dirt on

4 Please note: when the term "good mind" is used as one of the names of Sky Woman's eldest grandson, it is the first letters of both "Good" and "Mind" that have been capitalized. However, when the term "good mind" refers to the Haudenosaunee philosophical and cultural concept often discussed by elders and scholars, it is spelled with lower case letters. This is because (a) the Haudenosaunee elders and scholars referring to the cultural concept of the good mind do not, themselves, consistently spell this concept with upper case letters, and (b) using lower case differentiates the concept of the good mind from Good Mind, Sky Woman's Grandson (i.e. "he who formed our bodies").

Turtle's back and dances it into what we now know alternately as Turtle Island and North America. Thus, Sky Woman, with the help of birds and water animals, initiates creation. Eventually, she gives birth to a daughter. As time moves on, this daughter marries and gives birth to twin boys. Her elder son creates human life (Porter *And Grandma* 40-91; Doxtator 29-41; Mohawk *Iroquois Creation* 35-52; Mitchell 7-8).

One of the elder grandson's many names is "Good Mind" ("The Environment" 47). Gansworth himself prefers this name to the others Sky Woman's eldest grandson adopts in his time on this earth:

> Sky Woman has the child she was pregnant with, a daughter, who then rapidly grows up, meets a man and becomes pregnant, herself, with twins. One twin is born the natural way, but the other, obstinate from the start, chooses to leave his mother's body from her armpit, killing her. They're known by various names in different versions, Earth Grasper and Flint, Creator and Warty One. I prefer to call them the Good Mind *(backlighting on ADULT MALE, standing, stage left)* and the Bad Mind *(backlighting on ADULT MALE, standing, stage right)*. (52-53)

To emphasize the importance of these names, Gansworth goes on to explain that "bad" does not mean evil but, rather, "unsettled and unsound" (53). The implication here is that Good Mind is good because his mind is settled, peaceful, healthy and sound. Gansworth's choice to provide a cultural definition of the "Bad Mind" twin is critical in understanding the importance of the concept of the good mind in Haudenosaunee culture because, as Gansworth points out at the end of his play, the creation story is about balance. "The Good Mind does not conquer the Bad Mind" (74); rather, their creative works complement each other, creating a world that is held in balance by what each has to offer.

Furthermore, as it relates to Haudenosaunee cultural identity, the concept of the good mind once again surfaces in Haudenosaunee traditional teachings when the Peacemaker comes and unites the original five nations of the Haudenosaunee into a confederacy. To establish this union under the Great Peace, the Peacemaker travels throughout the lands of the Mohawk, Oneida, Onondaga, Cayuga and Seneca nations, working with people individually to heal them of their grief, anger, hatred and jealousy. He undertakes this journey during an era commonly referred to by Haudenosaunee storytellers as the "dark and troubled times" because of the excessive bloodshed and violence characterizing nearly all human relations in these lands. It is this violence and bloodshed that twisted the minds of the people, preventing them from using rational, peaceful, healthy minds to create stable communities. Little by little, the Peacemaker works through the power of the good mind to heal the minds of the people. It is also through the power of the good mind that the Peacemaker successfully re-creates the unstable battlegrounds of the nations into a stable landscape capable of sustaining the Great Law of Peace, which binds the original five Haudenosaunee Nations (and later Tuscarora) into what is now known as the Iroquois Confederacy or the Six Nations Confederacy (Williams "Kayanerenko:wa" 255-57). To ensure its survival, those men who are called upon to serve as chiefs are described as "he of the Good Mind" (Porter "Men Who Are of the Good Mind" 17); likewise, those women who are raised up as clan mothers are told they must always "carry Good Minds" in the work with the people of their clans (Thomas "Clan Mother's Role").[5]

In recent years, Haudenosaunee people have studied this philosophical concept to understand how it forms the basis of Haudenosaunee cultural identity. The editors of *The Great Peace: The Gathering of the Good Minds* CD-ROM spent quite a bit of time outlining the characteristics of the good mind for the purpose of teaching students of all age groups [using the CD-ROM as an educational tool] that true peace, respect, and co-existence as families, clans, and nations within the Confederacy depend upon people living and behaving through the power of the good mind. Thus, as Beverly Jacobs (Mohawk, Bear Clan) explained to the Canadian Law Commission, "within all of these teachings, it is said and understood that each person must have ganikwi:yo (a good mind)" (Jacobs 4). Paul Williams (Onondaga, Six Nations) tells us that "a good mind will seek peace, create peace, and will do so by uniting with other Good Minds" because power can be found in the unifying of good minds (Williams "The Great Law" 10; Williams "Kayanerenko:wa: The Great Law of Peace" 256-57). In this way, the good mind is both a healthy, peaceful, rational mind and the process by which people's minds are healed from those losses, emotions and events that make it difficult for people to think rationally. It is one of the hallmarks of Haudenosaunee cultural identity (Mohawk "Origins of Iroquois Political Thought" 218-28).

As *Re-Creation Story* opens, the audience soon becomes aware that the main character, Eric Gansworth, is riddled with anxiety, fear, doubt and resentment because his self-esteem and self-identity have taken a direct hit. First and foremost, as he indicated in his fateful reply to the listserv reminder, Gansworth, as a Haudenosaunee man, feels his knowledge of the Haudenosaunee creation story is completely inadequate (15). Gansworth also discovers, through his own research,

[5] "Clan Mother's Role," authored by the late Cayuga Chief Jacob Thomas, was published on the Web by a now defunct website Tuscaroras.com: http://tuscaroras.com/jtlc/The_Great_Law/chief_jake_thomas_clanmothers_role_pg2.html.

how little he knows about this story. As he learned just weeks before his talk, there is a "whole extended first section" of creation that takes place in Sky World. In fact, so much happens in Sky World that Gansworth fails to recognize Sky Woman until she falls through the hole in Sky World, which is made by uprooting the tree that provides light to all of Sky World (10-11). Later in the play, Gansworth tells his audience that he does not remember when he first heard the creation story. In fact, while Gansworth's visual art often includes well-known highlights from the creation account, his first vivid memory of hearing the creation story in any actual detail occurred at an installation of his artwork in a museum in western New York (35). For the first time, he realized his knowledge of Haudenosaunee creation is woefully inadequate. By the time Gansworth finished presenting the creation story at the Canisius College speaker series, he reconfirmed for himself that he possessed too little knowledge and understanding of this story to take on the role of a traditional Haudenosaunee storyteller, or to present creation to a listening audience of any size for any length of time. In *Re-Creation Story,* when Gansworth finishes his initial presentation with far too much time to spare, he looks to his colleague and complains:

> Shit! You know? I knew I wasn't going to fill three hours with that story. (to ADULT MALE) Didn't I tell you? And I knew there would be people from home in the audience. This is all I need, to be embarrassed by my inadequacies in public. (9)

Students and faculty seated in the classroom for the lecture series aggravated Gansworth's taking sense of Indianness by agreeing that his presentation "didn't take anywhere near three hours" as it was supposed to (30).

In addition, as becomes painfully obvious in a conversation with his niece, sitting in the latter's kitchen, Gansworth feels his skill as an oral storyteller, in the traditional Haudenosaunee sense of the word, leaves much to be desired:

> ADULT FEMALE: So I hear you're telling the Creation story on your birthday. I'm gonna try to come.
>
> ERIC: No, don't come. It's going to be terrible. I don't know how to tell the Creation story in a formal way.
>
> ADULT FEMALE: Yeah, I hear that, too. (19)

To make matters worse, Gansworth's niece confirms what a couple of Haudenosaunee elders have already told him: that he lacks the requisite skills to tell this story in public. Gansworth is asked throughout the play why his colleague thinks he can tell this story (13). Gansworth's niece wonders what made his colleague think "you could tell the Creation story anyway?" After all, "[i]t's not like anyone can tell it. There are kind of specific skills involved. Don't they know that?" (22). An unidentified female elder drives this point home by asking Gansworth, "Does [your colleague] not know you at all? All it would take is one look at your whacked out gasoline-pump-hoses-making-love-to-Indian-women paintings to know you're working with different tools than the usual Indian bears, deer, wolves, lightening, etc." (29). Gansworth's returns to these conversations throughout the play serve one purpose—to send him spiraling downward into a dark chasm of anxiety, self-doubt and resentment.

The night of the presentation, Gansworth's character is described as "*shuffling papers, nervous, restless, looking at paper, looking at audience, looking away, going back to look at papers again* [...] *trying unsuccessfully to sound like a professional storyteller*" (6). It is a long process of re-creation before Eric Gansworth, the character *and* the playwright, comes to understand what many of the elders from his culture already know: all traditional storytellers tell the creation story "in a way that reflects [their] limitations and personal experiences" (39). A storyteller can only see creation as he/she sees it because "the story tells as many things about us individually and our time as it does about its own timeless events" (39). This simple truth, in fact, mirrors my own experience hearing Mohawk elder Tom Porter tell the creation story several times at Kanatsiohereke, a small Mohawk community in upstate New York. When Tom talks about creation, specifically when he talks about Grandmother Moon—who in some versions is the Twins' mother and in other versions is Sky Woman herself—he smiles playfully, describing, her to his audience in much the same way his elders described the Moon to him: as a beautiful woman who likes to "get dolled up when she goes out at night," by adorning herself with the stars, like human women adorn themselves with diamond jewelry. Not only does Tom Porter's portrayal of Grandmother Moon tell us something about him and his deep abiding respect for his elders and for the Moon, but it also presents a visual image to his own audience that helps them relate to the Moon on a deeper level. In this way, the job of a traditional storyteller, entrusted with the telling of the creation story, becomes clear. The traditional storyteller must know the story well enough to weave his/her own experiences into the telling in a way that makes the story and its lessons relevant to his/her audience.

In this way, the traditional storyteller, according to yet another elder who advises Gansworth in *Re-Creation Story*, is someone who is able to tell the story in a way that ensures the story is re-created—"born anew in familiarity" (39). This is so because the story that is told and the story that is heard "are almost never the same thing" (39). To complicate matters further, the way a story is told, details included and omitted, are dictated by the people sitting in your audience listening to you telling the story: "Maybe you know that you've told a story to two different people, two different ways, two minutes apart" (39). It is this realization of how storytelling functions for both the storyteller and his/her audience, and how stories are recreated in the telling, and, specifically, how the re-telling in this specific play functions to restore Gansworth to a good mind, a mind that is at peace and able to rationally understand, accept and embrace the very processes of re-creation. These processes constitute a transformation for Gansworth, which takes place while he re-tells the story up to four times during the production of *Re-Creation Story.* Starting as a contemporary novelist, poet, playwright and visual artist, he becomes a traditional Haudenosaunee storyteller, entrusted with carrying forward the traditional teaching of Haudenosaunee creation in a way that ensures this story will remain relevant, for the benefit of future generations:

> Our Creation Story asks us merely to be thankful for what we have, to acknowledge those who have provided the opportunity to live a good life on earth, and to accept that at some point, it will come to an end. The version I gravitate to is the one in which things balance out. The Good Mind does not conquer the Bad mind […] They coexist, singer, and listener, writer and reader, weed and flower […] Do we become new people when our lives change that dramatically, when we leave the role of child behind? Sure […] the shoot within these seeds will crack their shells and fall or push themselves into new worlds [like Sky Woman falling through the hole in Sky World to the water world below], offering the possibility, even as we acknowledge loss, that there will be more beauty for which we should be thankful. N'yoh. (74-76)

Thus, by juxtaposing the unstable spaces of Haudenosaunee creation with a dramatically staged college classroom, Onondaga playwright Eric Gansworth's *Re-Creation Story*, dramatizes the re-creation of the ecology of the Haudenosaunee cultural landscape. In the ecology of this landscape, Gansworth dramatizes how, like seeds, humans initiate their own re-creation by breaking through their seed coats and pushing themselves through these dark holes into unstable, water-like "new" worlds—such as the classroom where the speaker series lecture is held, or the stage where the first play of a fledgling playwright is produced. Here, in these new and unstable worlds, Gansworth further dramatizes the ecology of the Haudenosaunee cultural landscapes by highlighting the importance of humanity's reciprocal relationships with other beings of creation who, like Turtle, may have sacrificed their lives, so that human life will continue. By so doing, Eric Gansworth dramatizes the ecology of Haudenosaunee creation in the following way: Good Mind and Bad Mind, weeds and flowers, Sky Woman and Turtle Island, are all reciprocally related to each other. They live together in a delicate state of balance that must be maintained through constant re-creation.

Works Cited

Doxtator, Deborah. "Godi'Nigoha' The Women's Mind and Seeing Through to the Land." *Godi'Nigoha' The Women's Mind.* Brantford: Woodland Cultural Center, 1997.

"The Environment." *The Great Peace: The Gathering of the Good Minds.* Brantford: Working World Training Centre Inc., 1998. CD-ROM.

Gansworth, Eric. *Re-Creation Story: A Play.* Niagara Falls, 2008. Message to Nicholle Dragone. 7 December. 2011. Email. [unpublished manuscript of the play produced in 2008].

The Great Peace: The Gathering of the Good Minds. Brantford: Working World Training Centre, Inc., 1998. CD-ROM.

Mitchell, Michael, ed. *Traditional Teachings.* Cornwall Island: North American Indian Traveling College, 1984.

Mohawk, John C. "Foreword." *Iroquois Creation Story: John Arthur Gibson and J.N.B Hewitt's Myth of the Earth Grasper.* Buffalo: Mohawk Publications, 2005. i-xix.

—. "Origins of Iroquois Political Thought." Ed. Joseph Bruchac. *New Voices from the Longhouse.* Greenfield Center: The Greenfield Review Press, 1989. 218-28.

—. *Iroquois Creation Story: John Arthur Gibson and J.N.B. Hewitt's Myth of the Earth Grasper.* Buffalo: Mohawk Publications, 2005.

Porter, Thomas. "Men Who Are of the Good Mind." Ed. Jose Barreiro. *Indian Roots of American Democracy.* Ithaca: Akwe:kon Press, 1992.

—. *And Grandma Said: Iroquois Teachings as Passed Down Through the Oral Tradition.* Bloomington: Xlibris Corp., 2008.

Thomas, Jacob. "Clan Mother's Role." *Tuscaroras.com* Web. 4 January 2001. http://tuscaroras.com/jtlc/The_Great_Law/chief_jake_thomas_clanmothers_role_pg2.html.

—. "The Great Law." *Tuscaroras.com* Web. 4 January 2001. http://tuscaroras.com/jtlc/The_Great_Law.

Williams, Paul. "The Great Law." Draft. 1 April, 2004. *"Indian" Indian Law Course No. 808.* Course Packet. Profs. E. Meidinger and J. Mohawk. Buffalo: Buffalo University School of Law. Spring 2005.

—. "Kayanerenko:wa The Great Law of Peace." Unpublished Manuscript. Middleport, Ontario, 2012.

Voicing Nature in the British Columbia Interior

A Place "Back in the Midst of Time" in Kevin Loring's *Where the Blood Mixes*[1]

Ginny Ratsoy

Thompson Rivers University

An ecocritical approach evolves "from the fundamental premise that human culture is connected to the physical world"; it "has one foot in literature and the other on the land," writes Cheryll Glotfelty—providing a solid base to explore a relationship that can be quite abstract (Glotfelty & Fromm xix). As Gabriella Giannachi and Nigel Stewart elucidate, conceptions of the relationship of nature to culture have been widely discussed in academia in the last half century, and are widely divergent—from Lyotard's assertion that nature must be interpreted from cultural parameters to Soper's claim that nature is independent and authentic (19). Raglon and Scholtmeijer illustrate the "nature as a category of language" concept in a concrete statement: "This makes nature a thoroughly human product, much like soap, cars and computers" (248). Soper's response to the concept of nature as a construct is a lucid declaration: "It is not language that has a hole in its ozone layer" (123). This essay views nature as tangible, genuine, and valid in its own right; furthermore, although an explicit argument for a more environmentally-friendly society is not its focus, it takes a dynamic medial perspective in line with Giannachi and Stewart's conception of nature as "a *process* of endless exchange and interactivity *between* the human and the other-than-human" (19). My ecocritical approach ascribes a fluid, interconnected relationship to the human-nature dynamic; furthermore, it highlights the connection between the natural and the spiritual: the fundamentally ecospiritual nature of the text under consideration is focal in this study.

---

1 I wish to acknowledge the research assistance of Patricia Johnson, Information Specialist, Gallagher Library, University of Calgary.

This essay examines nature/human interdependence in Aboriginal playwright Kevin Loring's first play. Into a contemporary setting, *Where the Blood Mixes* (2009) incorporates traditional N'lakap'mux (Interior Salish of British Columbia) conceptions of the interconnectedness of time, place, space, and being. These beliefs are perhaps best encapsulated by Nancy Turner, Marianne Boelscher Ignace and Ronald Ignace in "Traditional Ecological Knowledge and Wisdom of Aboriginal Peoples in British Columbia":

> The environment is seen as a whole; all the parts are interconnected in a seamless web of causes and effects, actions and outcomes, behaviors and consequences. People, animals, plants, natural objects, and supernatural entities are not separate and distinct. Rather, they are all linked to each other and to the places where they reside through cultural traditions and interactive, reciprocal relationships. Because of the integration of the secular with the spiritual, of the past with the present, and of all parts of the living universe, people have a sense of the spiritual and practical respect for their lands, waters, and all the environmental components that they recognize. (1279)

In this conception, environment—for both its physical and sacred properties—is integral to and inseparable from individual and collective existence.

Thus, this essay eschews the dualism on which much of Western history is based, an oversimplification that underpins imperialism and, as Lawrence Coupe details, often places nature in opposition and subordinate to culture, reason, mind, freedom, and civilization (119-20). As *Where the Blood Mixes* reveals the effects of particular aspects of colonization, it is important to recognize that colonization emerged out of dualism and cultural and ecological difference and to acknowledge Huggan and Tiffin's assertion that anthropocentrism and Eurocentrism are "inseparable" (5): the lands, people and animals that inhabited space colonized by Europeans were firmly placed in the subordinate position in this duality.

That the arts in the West—and theatre in particular—have been complicit in the separation and subordination of nature has also been widely discussed. Arons and May, for example, maintain that theatre's essence—"human conflict in the context of human institutions" (1)—subordinates nature to society. Una Chaudhuri, examining nineteenth- and early twentieth-century theatre specifically, sees naturalism and realism as "proffering a wholly social account of human life" ("Fish" 24). While naturalist dramatists perceived the environment as deterministic, she says, environment was almost entirely socially defined; realist dramatists made shadowy use of natural elements—reducing the animate to the inanimate. The idea of place in late twentieth-century theatre, Chaudhuri states elsewhere, undergoes a transformation in "dramatic structure as well as stage place" that is manifested primarily as "a departure from the 'solid state' world of naturalism, as well as from its logic of total visibility" (*Staging Place* xiii). In Canada, Aboriginal dramatists have made distinctive contributions to these theatrical departures. Rob Appleford asserts that the greatest strength of the body of work by Aboriginal dramatists in Canada "lies in the delight of the unforeseen" in "both the unpredictability of the narrative's focus, and the theatrical means through which this focus is effectively rendered" (viii). Loring's play eschews the solidity and transparency of naturalism in a shifting narrative focus. Furthermore, the means by which Loring renders nature not only as a sacred force inseparable from humans but also as a device to disrupt Western notions of temporal distinctions is, in my mind, one of the most striking and delightful aspects of his work.

Where The Blood Mixes also follows important conventions of what Thomas King calls "associational literature" in his 1990 essay "Godzilla vs. Post-Colonial" (where he opposes the imposition of the term "post-colonial" on contemporary Aboriginal writing): "Associational literature leans towards the group rather than the single, isolated character, creating a fiction that de-values heroes and villains in favour of members of a community, a fiction which eschews judgements and conclusions" (246). The play eludes classification as comedy or tragedy, while incorporating strong elements of both; it also places emphasis on the group and defies easy categorization. While it must be acknowledged that these are characteristics of a not insubstantial body of much late twentieth- and early twenty-first-century drama, Aboriginal or otherwise, it must also be recognized that the approaches of many contemporary Indigenous writers, Loring included, are in significant measure legacies of traditional Aboriginal storytelling—episodic, communally focused, and circular in narrative style. King further characterizes associational literature as keeping a distance and hiding certain aspects from its non-Native readers: "Non-Natives may, as readers, come to an association with these communities, but they remain, always, outsiders" (247). Although Loring's play is thoroughgoing in its exploration and revelation of, for example, some aspects of character, it is resistant to full natural/cultural/spiritual revelation, leaving the reader who lacks experiential knowledge of its cultural base in a liminal position. Furthermore, the playwright uses space to conflate time, specifically in ways that subordinate the latter to the former and eschew linear temporal designations. Temporal elision through elements of nature intensifies the effects of colonial injustices, underscores the intimate connections between the land and its inhabitants, and enforces traditional Aboriginal views about time as circular rather than linear: the people are *in* the land; the past is *in* the present—and this is also very much the case with *Where the Blood Mixes.*

Where the Blood Mixes is imbedded in the natural landscape of Lytton. About 150 kilometres south of Kamloops, Lytton (population circa 300) shares with it a role in one of the world's largest salmon runs. Adult salmon leave the ocean in the fall, enter the Fraser River at its junction with the South Thompson at Lytton and then progress up the South Thompson, ending their journey east of Kamloops at the Adams River, where they spawn and die. Lytton is also known for one of the largest collections of pictographs in Canada, in and around the nearby Stein Valley N'lakap'mux Heritage Park. The village acquired its second English name during the gold rush of 1858. The Colonial Secretary of England, Lord Edward George Earle Bulwer-Lytton, was also a popular writer in his time, but is most remembered for the immortal line, "It was a dark and stormy night" (*Lytton River Festival*), which has inspired writing contests (at San Jose State University and Kamloops's Thompson Rivers University, for example) that encourage deliberately florid language. The village's first English name, The Forks, was designated following Simon Fraser's 1808 visit in search of a route to the Pacific (*Lytton River Festival*).

However, the town's N'lakap'mux name, Kumsheen, is many thousands of years older than the two European designations, and its connections to landscape are deep. "Where the blood mixes" is a translation of "Kumsheen," which, according to Loring, refers to the confluence of the Thompson and Fraser Rivers, the blood-red colour of the confluence during the salmon run, and, most significantly, "the place inside the heart where the blood mixes" (93). The last meaning, Loring says, alludes to a N'lakap'mux story: the intestines of Coyote, the loser in an epic battle with the Transformer, imbue the hills where they landed with their distinctive colour, and Coyote's heart fell and remains at that confluence (93-94). In a review of the Tarragon Theatre production of *Where the Blood Mixes, Toronto Star*'s Robert Crew pronounces the play as having "a heart as big and as beautiful as the mountains of British Columbia," eloquently touching the surface of the work's ecospirituality (Crew). The landscape connects nature to culture and spirituality, and characters to both—as well as each other.

Where the Blood Mixes also scrutinizes historical injustices whose effects are ongoing. The residential school system, from the 1830s (in pre-confederation Canada) until well past the time the decision was made to close them (according to J. R. Miller, the last was closed in Saskatchewan, in 1996), would prove one of the most devastating effects of colonization. At these schools, food was often scarce and poor and pedagogy deficient; the curriculum "reflected an alien culture" (Miller). About St. George's School in Lytton, former student Nathan Spinks recounts, "You went only two and a half hours a day for academic instruction and two and a half hours a day for manual labour training (framing, carpentry, and boiler room for boys; cooking, serving, and general housecleaning for girls)" (137). As the "statement of reconciliation" issued by the Canadian government on January 7, 1998 acknowledged:

> This system separated many children from their families and communities and prevented them from speaking their own languages and from learning about their heritage and cultures. In the worst cases, it left legacies of personal pain and distress that continue to reverberate in Aboriginal communities to this day. Tragically, some children were the victims of physical and sexual abuse. (*Aboriginal Affairs*)

Despite this and subsequent reconciliation initiatives, and, in 2008, an apology by the Canadian Prime Minister acknowledging the "assimilation policy on which the schools were established was 'wrong, has caused great harm, and has no place in our country'" (qtd. in Miller), this government practice has had enduring effects, as has a second one Loring's play references—at its height in the 1960s—of "scooping" children from their homes to be placed in non-Aboriginal foster homes.[2] In both cases, familial and cultural damage was deep and lasting, and in both cases a profound disconnection from the traditional ecological knowledge of communities resulted.

Where the Blood Mixes explores a path to healing through community and environment. Specifically, in its characterization and dramatic trajectory, the play incorporates seemingly disparate, even incongruous, elements into a whole that embodies the cultural heritage of the N'lakap'mux people. The mundane is side by side with the sacred, the past and present are virtually inseparable, and land, water, and creatures interact with, act upon, and profoundly influence character and action. Certainly, as Turner, Boelscher Ignace and Ignace put it, the traditional wisdom of the area that "people, animals, plants, natural objects, and supernatural entities are not separate and distinct" (1279) is evident throughout. Populated by not only corporeal characters but also ghosts and elements of the (past and present) natural world, Loring's play, through a theatrical amalgamation, posits that a community that has been rent asunder by various projects of colonization can only begin to nurse itself back to health by confronting the damaging effects those projects have had in order to reconnect to and reclaim that wisdom. Colonization has produced a communal and individual disconnect; reaching to a more distant, pre-contact ecospirituality is the key to integration.

2 "Moonlodge" (1990) by Margo Kane (Cree/ Salteaux) is perhaps the most well-known Aboriginal play in Canada to address the practice of scooping.

Each of the four Aboriginal characters in *Where the Blood Mixes* remains profoundly affected by the residential school in multiple ways. In addition to suffering individually from the experience, the three who attended residential school share a collective guilt: they are haunted by the absence of Anna, who, because of her own residential school experience, took her life by jumping from the town bridge. Her partner, Floyd, has been unable to move on, his grief compounded because their daughter, Christine, was adopted out in the city after Anna's death. Mooch, Floyd's closest friend, who was physically present at Anna's death and blames himself for not preventing it, also suffers the ongoing psychological effects of his sexual abuse at the school. Mooch is at once the most profoundly affected by the residential school experience and the male character most clearly associated with attempts to reach the natural world and (inseparable) cultural past of the N'lakap'mux people. The scars are most visible in the two male characters, who drink excessively. Mooch's partner, June, has come to terms with her residential school experience, in large measure because of her communal relationship with the natural world, as I shall discuss. However, the suicide of her best friend haunts June, who feels guilty for not having prevented Anna's death and not having adopted Christine. Although Christine, now a young adult and an artist, has been satisfied with her adopted family, she feels compelled to return home. Her arrival—anticipated and actual—precipitates much of the play's action, dream sequences, reflections, and remembering. Examining four Cree plays, Mary Blackstone asserts that because "aboriginal concepts of time [are] cyclical rather than linear or progressive, we must situate memory as a liminal site in which past, present and future intersect in the ongoing negotiation of meaning and identity" (180). Blackstone's generalization holds true here, where the present overlaps with several pasts: Anna's suicide, Christine's departure, the past of the residential school and its subsequent effects, and the pre-contact past, for example. Natural elements are the adhesive that conjoins those pasts to one another and to the present and future.

The present action of the play unfolds during the time the government announces compensation for the victims of the residential school. The settings are largely communal—group spaces—most notably two distinctly different watering holes. The town bar, which features the play's only non-Aboriginal character, bartender George, is the primary indoor setting. Although his name might recall the residential school in Lytton, George functions, as Robert Benz, who played the character in the Western Canada Theatre production of the play, says, as primarily a witness and a substitute for the theatre audience (Benz). Outdoor settings predominate, especially various spots on the river. Much of the effect of the play depends on Loring's deft handling of the two disparate settings.

The bar—a static limbo where Floyd and Mooch are suspended, often arguing or physically fighting with each other, George, or June— contrasts with outdoor scenes, which often conjure up a multi-layered past both traumatic and magical and are associated with movement towards healing. Outdoors, characters act either individually in productive steps toward self healing or in concert towards a common good. Natural elements are more than a backdrop to these actions; in fact, as they connect characters to the spiritual, they often seem to precipitate them. Directive transitions are few and far between: setting and time shifts are usually indicated by alterations in presentational modes. Bar scenes are largely in realist/naturalist mode; outdoor scenes are expressionistic. Only two scenes are in a private setting—Floyd's home, as he prepares for and contends with Christine's arrival. The dominance of communal space complements the roughly equivalent significance placed on the four Indigenous characters.

While the Coyote story is narrated by June near the play's end (as well as being explained by Loring in an afterword), the other prominent element of land is conveyed through stage directions—visually rather than orally. Pictographs—visual clues and enhancements—are strategically placed in relation to Christine. According to the late N'lakap'mux elder, healer, and teacher Annie Zetco York, traditionally the elders encouraged the youth in a solo rite of passage entailing separation and solitude followed by a return. This spiritual quest was a means of selecting candidates for shamanism (ix). York narrates, "They go up there [in the mountains], and they sleep, and this dream tells them. Then he writes his dream on the rock. That's left there forever" (xv). The pictographs were seen as both writing and drawing: the N'lakap'mux people call them Ts'ets'ékw, which has been translated as both writing and "a picture of any kind" (York, Daly and Arnett 3). The medium is powdered ochre mixed with water, resulting in a red that York says represents life and protection (4). The sites of these writings were usually carefully selected: the canvases or tablets are spots where nature's mysterious forces were thought to be in great abundance, chosen in hope that this power would be transmitted to the questor. The subjects of these records could be the questor's most important visions; his/her wishes; positive signs he/she had seen, or his/her adventures, especially those related to animals (York, Daly and Arnett 5-6). Occasionally, other community members would create the writings as protection from evil or as a historical record of such events as battles or the deaths of chiefs (York, Daly and Arnett 6-7). A visual story, the pictograph confirms inhabitation, imbeds the creator's identity within place, and produces a record of culture and spirituality, as well as connecting the questor—through nature—to spiritual realms. In a traditionally oral culture, it assumes a particularly significant role as document.

Not only does the association of the pictographs with Christine reinforce her vocation as an artist (she creates spoken word poetry as well as painting and drawing), but it also provides a visual reminder that her place of origin cannot be denied. Like coyote's intestines on the mountain, the pictographs are enduring evidence of the interconnectivity of identity (individual and cultural) and landscape. Among the opening words of the book are these: "Enter young native woman wearing a simple white dress.... Projected onto the woman is a pictograph, revealed in the shimmering light. She radiates a ghostly aura" (11). This ambiguous description (it seems to at once refer to Christine, Anna, and an Aboriginal Everywoman) not only alludes to Christine's striking physical resemblance to Anna (who is still spiritually present) and the haunting effect of Christine's own absence, but it is also a visual metaphor that evokes Christine's and her late mother's spiritual powers, as well as Christine's own shamanism, which bears out over the course of the play.[3] Subsequent references to the pictograph are also visual: just after an expressionistic scene, entitled "Chorus of the Lost" where Christine is connected (by a visual of a multitude of crumpled letters floating in the river) with others in various pasts who were displaced and just before an important fishing scene, pictograph salmon briefly appear and disappear; again, nature and art are graphically connected to identity. At Floyd and Christine's reconciliation near the play's end, Christine's childhood drawings remain on his walls and are superimposed upon by "house, fish, people"—a pictograph collage—a tableau which, strikingly, Christine herself briefly enters. Thus, the relationship between the pictographs and the character is explicit in a graphic reinforcement of plot elements. Like the art on the landscape, the pictographs in the play are reminders of the importance of Christine's (birth)place and immanence. The fluidity of the relationship between nature, spirituality, and human is most visually arresting in the pictograph scenes.

Ecospirituality is established at the play's outset, by the river, which introduces the play. The connection between the landscape and the character is explicit and intricate, as Christine utters the first words:

> I was born in the heart.
> I was born in the deepest part.
> In the middle of it all, I was born.
> In the place where the rhythm beats,

---

3 The Western Canada Theatre production of the play elected to understand Christine as the native woman—an entirely justifiable decision, as Loring designates the first speech (which immediately follows the stage directions) to Christine. However, the ambiguity must be noted, and I am indebted to my English 1140 student Katarina DeSimone for bringing it to my attention.

> Deep inside my mother,
> Where the rivers meet.
> My father dreamt me there.
> Where blood mixes with blood and the sturgeon waits,
> And the wind sings the songs of the dead. (11)

As Christine's appearance introduced the pictographs, her initial words introduce the other significant recurring natural images. Her voice announces the play's genesis in place. Here, not only is the river the metaphor for life, but also the wind immortalizes the dead. The sturgeon reappears in different contexts throughout the play (as do salmon). Like the pictograph-transformed mountains, the other natural elements are contextualized to varying degrees in ways that place them—rather than in a separated, undisturbed natural world—in an intricate and sacred web with humanity.

In the pictographs, visible interactions between landscape and society—at once personal and communal and contemporary and historical—with lasting consequences take place. The play develops and articulates these connections quite patently. However, there is a level of the natural world where that intersection is slippery and elusive: the waters where Coyote's heart landed. *Where the Blood Mixes* culminates with the "salmon swimming home, pulled by an overwhelming need to return to the place of their birth" (90). Although this image most obviously parallels Christine's situation, it rings true for the other characters on the singing bridge; June, for example, has had profound connections to the salmon, in both the past and present, which catalyze her spiritual growth, and Floyd's and Mooch's journeys are in other ways linked to fish. However, deeper in the river into which Anna disappeared lies a more mysterious force to which the play's last words are devoted: "Still further we descend. In the dark light of a deep grey pool the ancient sturgeon waits, her long sensuous ghost-grey body gently swaying in the current below" (90). As in the beginning, so, too, the sturgeon waits at the end.

In the midst of the play, the sturgeon has various complex implications. The salmon pictograph's brief appearance immediately subsequent to the "Chorus of the Lost" is a visual transition to a fishing episode featuring Mooch and Floyd, who, in mock heroic mode, engage in a contest of masculinity to determine which of them gets to fish in a favoured spot. Mooch's victory is short lived, as the arm wrestling induces cramps. No winners emerge in this contest of territoriality, which immediately prefigures two fish stories—one in the past and one in the present—that upturn and deflate Western heroic narratives. Sturgeons, toothless bottom feeders who can survive in both fresh and salt water, are rife with mythic possibilities: the white sturgeon of the Fraser River can exceed six meters in length, 600 kilograms in weight and 150 years in age, and, according to the Fraser River Sturgeon Conservation Society, is "the last truly wild population of this species in the world, the largest and longest-lived species of fresh water fish" on the continent, and "virtually unchanged for 65 million years… a survivor from the age of dinosaurs" (*Fraser River Sturgeon Conservation Society*). Mooch recollects in considerable detail the past "sturgeon that got away" story: enlisted by his grandfather to assist in pulling in the fish, which Mooch remembers as "a sea monster" and "a great big boney-grey crocodile-looking dinosaur fish" (38) in a complex, lengthy endeavour also involving a horse, the ten-year old, fearful of being eaten alive, ran away instead, as the sturgeon subsequently did. Despite the heroic lead up, the failure is not made much of: although his papa would not speak to him for a month, when he does he laughs extensively, telling Mooch "You make good bait" (39). Thus, time transmutes failed heroics into comedy. Soon after this recounting, Floyd, unaided by Mooch, hooks a sturgeon, but the line snaps and the sturgeon, again, swims downstream. This episode is presented briefly and unceremoniously. While the present pales beside the past here, what the present experience evokes is significant—both Mooch's past story and Floyd's missed opportunities. Resigned to his failure, Floyd makes a statement that prefigures future encounters with the sturgeon: "Next time, next time I'll be ready" (42). Dialogue and a flashback involving Christine are precipitated by the fish stories: after Mooch's tale, Floyd informs Mooch of Christine's desire to visit; immediately after Floyd's sturgeon encounter, Floyd and the child Christine are in a dialogue that should have served as a warning that Christine would be taken from him; and, immediately following that is Christine and Floyd's recent phone conversation about her imminent visit. Clearly, the sturgeon incident is a reminder of similar examples of a lack of preparedness on Floyd's part, and, once again, a natural element links past and present. The sturgeon stories, thus, become inextricably linked to central plot elements.

In addition to paralleling similar human (in)action and indicating the importance of cooperation rather than contest, the sturgeon stories may be read as meta-theatrical comments on the trajectory of the play itself—do not expect clear-cut endings of solo human heroics here—as they reveal the inscrutability, evasiveness, and indomitability of nature. Despite lengthy and elaborate preparations, the past attempt to catch the sturgeon was a failure; the comparatively brief and spontaneous attempt on Floyd's part meets with equal results. The first story follows a comic trajectory, but the second is neither comic nor tragic: it is resigned, or more accurately, accepting. The fish stories, respectively, invert, and, more complexly, subvert the conventional Western human-nature contest.

As the absurd arm wrestling over territoriality produces no real winner, the competition with the fish does not display man's dominion over nature. The sturgeon, presumably unperturbed, swims away, elusive—a force not to be reckoned with, and not fully understood.

The role of the sturgeon becomes murkier as the play progresses. After a bar scene in which June confronts Mooch about money he stole from her purse, the confrontation culminating in her hitting him, Floyd, intoxicated, gets waylaid on his way home and finds himself paralyzed on railway tracks, even as a train bears down on him. His deliverance comes in the form of a vision of "a young Native woman" (51) presenting him with an unidentified bundle that soon unravels and the woman then takes from him as she departs. Echoing the play's beginning, this expressionistic conjoining of Anna and Christine (and an Aboriginal Everywoman) reflects the vital, fragile and elusive role of the women in Floyd's life: they are his salvation, but they, too, got away from him. The scene also underscores Floyd's disconnection from the natural and spiritual worlds, as does his subsequent discussion about the sturgeon's life cycle with Mooch, who suddenly materializes. Floyd espouses a secular, material view; however, Mooch sees a cyclical pattern of life and death in which the sturgeon plays a seminal role, a spiritual belief passed down from his grandfather. In this view, the afterlife takes place underwater, and Mooch makes it there in an observer role: he has confronted death, possibly contemplated suicide, but not taken the plunge. Discussing Anna's drowning and her body's disappearance, Mooch speculates that the sturgeon eats the bodies of "the ones that commit" (53) and thus instigates a cycle of spiritual immortality: the sturgeon eats their bodies; the sturgeon, in turn, gets eaten by humans, and, in that process, the drowned human lives on in living beings. The images of immortality and incarnation not only connect human to human, but also, of course, human to nature, in an ongoing process that is circular and, ultimately, unknowable—dependent on faith. After a disdainful rebuke from Floyd, who exits the river scene, Mooch dives into the river and emerges at the bar. As well as signifying Floyd's detachment from the traditional ecological beliefs of his people and Mooch's flirtation with suicide and affinity with the spiritual world, this visual expressionism also prefigures the play's final scenes, which take place on the bridge over the river.

The wind is connected throughout the play to a man-made structure—the bridge, which serves as both a connection and a separation: it was to the middle of that bridge that the child Mooch made it before he was caught escaping from the residential school, only to be returned by police to a waiting priest who subsequently "Beat me, starved me. Fucked me" (59). Throughout the play, the bridge is a crossroads in various literal and metaphoric ways. Repeatedly referred to as "the singing bridge," it is the wind's vehicle to recount the past. The day after Floyd has refused Christine's entreaties to re-establish their relationship, June and Christine walk along the bridge, which, through June, narrates history—the history of the name Kumsheen, traditional stories such as that of Coyote's mark on the mountain, and the inseparable personal history of June herself—histories Christine needs to know to reconnect with her community. Christine herself recalls reading about a N'lakap'mux belief that the lands of the living and dead are separated by a river and connected by a slippery log; in this vision of the vulnerability and tenuousness of life, the bridge functions as a stand in for the log. Further connections among nature, spirituality, and history occur when June recalls her own epiphany and transformation from alcoholism—precipitated by her sight of fish jumping into the lake—which in turn induced a flood of memories, including of her grandmother's lullabies. In the past, nature stimulated memories of ancestry and youth that lead to psychic healing. In the present, nature functions similarly: as June confesses her guilt about both Anna's death and Christine's removal from the community, she sees the salmon in the eddy and, in one of several examples of code switching in the play, then says a N'lakap'mux prayer of thanks—a tribute to the salmons' role as providers of nourishment and to the psychological life they restored in her, as well as an illustration of associational literature's propensity to include only to a point. Christine turns to Floyd, who has been sitting down river, mute, throughout this scene, and explains why she has returned home: "I couldn't move forward without knowing where I came from" (83). The two reconcile, with Floyd acknowledging "this is your place, too, eh" (83), and recounting the details of Anna's death. Like June, Floyd has now "sung" his song to Christine. Their final words reinforce Christine's reunion with home and family and reconnect Floyd to nature: "Christine: 'You smell like home'. Floyd: 'I stink like fish'" (87). In their reunion, both Floyd and Christine return home—a return abetted by June's recollection of her own spiritual "return"—itself caused by her reconnection to the salmon. The ecospiritual elements are patent: June is the human bridge, facilitated in that role by the physical sight of the salmon, which catalyses a spiritual recognition. The interconnectedness of which Turner, Boelscher Ignace, and Ignace write is graphically illustrated in the bridge scene, where nature is very much a dynamic, causal force.

The final scene sees June and Mooch in the middle of the singing bridge. June thinks Mooch may be contemplating jumping, as he recalls Anna's jump in considerable detail, concluding, "When she hit she never came up. She just disappeared" (89). However, Mooch, in earlier bar scenes, has talked about his residential school experiences and relived Anna's death scene by mistaking Christine for Anna and apologizing for his inaction, while holding Christine back (a powerful blurring of time, space, and person that indicates he has to some extent come to terms with his past). As significantly, he has held on to the beliefs in the spiritual significance of the salmon that his grandfather passed on to him. After verbalizing the trauma that witnessing Anna's death caused him, Mooch renounces his nickname, and June complies with his desire to assume his true identity as Edgar (89). The last words, which belong to Edgar and appear to reaffirm that his choice is life over death, reinforce the symbolism of the bridge as a crossroads between, among other things, the past and the present: "I cross this bridge every day" (89). These words are profound in their simplicity—and in their ramifications for all of the characters. The loss with which they are faced is mitigated by Christine's return—she does provide them with "a second chance" as June has earlier told Floyd and Mooch she would—and the three of them take advantage of that second chance, as they offer apologies to her and—through her—to Anna. In various significant ways, the three reconnect with the traditional ecological beliefs of their people, as evidenced by their respective connections to the salmon. Nonetheless, their experiences have shaped them and do not vanish in their transformation: the singing bridge over the river is a tangible reminder of the deep and mysterious role of the sturgeon and the unknowability of the whereabouts of Anna's body.

Tomson Highway has articulated the powerful historical relationship of Indigenous peoples to the landscape in an essay entitled "On Native Mythology" in which he describes that mythology as possessing "symbolism [...] as basic and direct as air" and as coming "from deep, deep within the flesh and blood of a people who have known this particular landscape since time immemorial, and who are so close to it they have become an integral part of it, like rock" (Appleford 2). Kevin Loring's *Where the Blood Mixes* graphically illustrates the extent and intimacy of this relationship.

Analyzing the use of nature in works by Gordimer, Hoban, and Kafka, Raglon and Scholtmeijer conclude that these writers have "confronted something that is resistant to human meaning" and acknowledge that, although nature in these works can be read in various psychological, symbolic, and metaphoric ways, their study placed them face to face with "nature's resistance to our stories" (261). To some extent, particularly in examination of the role of the sturgeon in Kevin Loring's *Where the Blood Mixes*, this writer has come to a similar position—a liminal state exacerbated, perhaps, by her lack of cultural experience. Although the physical world is the play's unifying force, nature—as it erases artificial western time delineations, unites history and place and connects individual and community—does not yield up all. Nature maintains some substantial level of mystery, which is perhaps in keeping with its spiritual role in the play. As Thomas King states about associational literature, "The access is not unlimited" (247). What the play does reveal, however, is key: nature is implicit in home, and home is key in individual and collective identity. Significantly, although Floyd and Mooch—and to a lesser extent June—are imprisoned by the past in Lytton, no dialogue is devoted to departure from the town. The cultural past of the colonized site is responsible for their problems, not the natural past; in fact, the healing they seek is to be achieved by *staying in* the place and reconnecting with its natural and spiritual elements, rather than escaping from it. Annie York, Richard Daly, and Chris Arnett indicate that the rock writings of the N'lakap'mux people are believed to have come from two different periods: the more recent the period immediately subsequent to contact and continuing for approximately forty years, and the earlier "back in the midst of time" (9), noting that "late period painting is sometime superimposed over early period paintings" (11). In its close connection to the historical cultural and spiritual world of the N'lakap'mux, Kevin Loring's *Where the Blood Mixes* is a contemporary equivalent of—a companion piece to—these rock writings: it is a twenty-first-century superimposition onto and mixing with the landscape through the medium of theatre that also incorporates earlier periods.

Huggan and Tiffin perceive Indigenous theatre as "[s]imultaneously a theatre of loss—land, language, cultural heritage—and one of resistance and reclamation" (121). In *Where the Blood Mixes,* Loring certainly evokes loss, primarily of individual identity and cultural heritage, but he also illustrates reclamation—through reconnection with cultural heritage and (inseparable) spiritual and natural elements of place. Huggan and Tiffin further characterize the Indigenous theatre stage as "a contested space in which conflicting spatial fantasies and histories are accumulated, and the land is revealed both as speaking subject and as object of discursive management and material control" (121). Through a tribute to traditional Aboriginal storytelling, as well as to the traditional ecological wisdom of the N'lakap'mux people, Loring creates a space in which nature interacts with, speaks through, and unifies space, time, and people.

Works Cited

Aboriginal Affairs and Northern Development Canada. "Notes for an Address by the Honorable Jane Stewart, Minister of Indian Affairs and Northern Development on the Occasion of the Unveiling of *Gathering Strength—Canada's Aboriginal Action Plan.*" 7 January 1998. Web. 22 August 2012.

Appleford, Rob. "Introduction." Ed. Rob Appleford. *Aboriginal Drama and Theatre*. Toronto: Playwrights Canada Press, 2005.

—. "Daniel David Moses: Ghost Writer with a Vengeance." Ed. Rob Appleford. *Aboriginal Drama and Theatre*. Toronto: Playwrights Canada Press, 2005: 150-65.

Arons, Wendy, and Theresa J. May, eds. *Readings in Performance and Ecology*. New York: Palgrave Macmillan, 2012.

Benz, Robert. "Acting in Western Canada Theatre's Production of *Where the Blood Mixes.*" Thompson Rivers University, Arts and Education Building, Kamloops, BC. 11 October 2012. Guest Lecture.

Blackstone, Mary. "Cultural Memory in Four Cree Plays." Ed. Marc Maufort and Caroline De Wagter. *Signatures of the Past: Cultural Memory in Contemporary Anglophone North American Drama.* Brussels: Peter Lang, 2008: 179-95.

Chaudhuri, Una. "There Must Be a Lot of Fish in That Lake: Toward an Ecological Theater." *Theater* 25.1 (Spring/Summer 1994): 23-31.

—. *Staging Place: The Geography of Modern Drama.* Ann Arbor: U of Michigan P, 2000.

Coupe, Lawrence, ed. *The Green Studies Reader: From Romanticism to Ecocriticism.* London: Routledge, 2000.

Crew, Robert. "*Where the Blood Mixes*: A Poignant Play with a Big, Beautiful Heart." *Toronto Star*. 24 April 2010. Web.

Fraser River Sturgeon Conservation Society. n.pag. n.d. Web. 3 August 2012.

Giannachi, Gabriella, and Nigel Stewart, eds. *Performing Nature: Explorations in Ecology and the Arts.* Bern: Peter Lang, 2005.

Glotfelty, Cheryll, and Harold Fromm, eds. *The Ecocriticism Reader. Landmarks in Literary Ecology*. Athens and London: The U of Georgia P, 1996.

Highway, Tomson. "On Native Mythology." Ed. Rob Appleford. *Aboriginal Drama and Theatre*. Toronto: Playwrights Canada Press, 2005. 2-10.

Huggan, Graham, and Helen Tiffin. *Postcolonial Ecocriticism: Literature, Animals, Environment.* London and New York: Routledge, 2010.

King, Thomas. "Godzilla vs. Post-Colonial." Ed. Ajay Heble, Donna Palmateer Pennee, and J.R. Struthers. *New Contexts of Canadian Criticism*. Peterborough, ON: Broadview Press, 1997. 241-48.

Loring, Kevin. *Where The Blood Mixes.* Vancouver: Talonbooks, 2009.

Lytton River Festival. n.pag. n.d. Web. 31 August 2012.

Miller, J.R. "Residential Schools." *The Canadian Encyclopedia.* n.d. Web. 23 August 2012.

Raglon, Rebecca, and Marian Scholtmeijer. "Heading Off the Track: Language, Literature, and Nature's Resistance to Narrative." Ed. Karla Armbruster and Kathleen R. Wallace. *Beyond Nature Writing: Expanding the Boundaries of Ecocriticism*. Charlottesville: UP of Virginia, 2001: 248-62.

Soper, Kate. "The Idea of Nature." Ed. Lawrence Coupe. *The Green Studies Reader: From Romanticism to Ecocriticism.* London: Routledge, 2000: 123-26.

Spinks, Nathan. "Old Ways." Ed. Darwin Hanna and Mamie Henry. *Our Tellings: Interior Salish Stories of the Nlaka'pamux People.* Vancouver: UBC Press, 1997. 137-39.

Stein Valley Nlaka'pamux Heritage Park. Web. 8 August 2012.

Turner, Nancy J., Marianne Boelscher Ignace and Ronald Ignace. "Traditional Ecological Knowledge and Wisdom of Aboriginal Peoples in British Columbia." *Ecological Applications* 10.5 (October 2000): 1275-87.

York, Annie, Richard Daly and Chris Arnett. *They Write Their Dreams on the Rock Forever: Rock Writings in the Stein River Valley of British Columbia.* Vancouver: Talonbooks, 1993.

Serving the Living Land

Place and Belonging in Australian Aboriginal Drama

Maryrose CASEY

Monash University

Introduction

European settler fear and distrust of the physical environment has been a notable feature in Australian literary writings since the nineteenth century. The main emphasis then was on what can be understood as a perceived hostile relationship between white settlers and the Australian landscape. Often in these works, Aboriginal people have been represented as another threatening element within this hostile environment. Examples include dramas such as Henrietta Drake-Brockman's *Men Without Wives* (1955) where the off-stage sounds of the Aboriginal people on the station are part of the intrinsic threats to European life and values (5). For the Anglo settlers the land was harsh and lonely, a threat that had to be managed.

In contrast, Indigenous Australians traditionally have a strong, positive and multilayered relationship with their historical homelands. For many, the processes of European colonisation that resulted in the removal of Indigenous peoples from their homelands complicates this relationship rather than lessening its importance. This essay focuses on the work of two Indigenous Australian writers, Wesley Enoch and David Milroy, specifically Enoch's *The Story of the Miracles at Cookie's Table* (2007) and *Black Medea* (2007) and Milroy's *Windmill Baby* (2007). These plays engage with issues of identity, place and belonging from different perspectives within a particular geographical location. My aim is to examine the specific relationships to the environment that is encapsulated in these texts. *Cookie's Table* is embedded in the tradition of Aboriginal people being born under a specific tree, one chosen for the gifts it can give such as strength and wisdom. Cookie's birth tree was cut down and made into a table that she and her family have managed to claim and pass down through generations. The play foregrounds the table as symbolic of country, belonging, the family and its future. In *Black Medea*, the relationship with country (as Aboriginal home lands are known) is personified through spirits of the family and land. The central theme is the emotional conflict caused by separation between people and country. *Windmill Baby* is located in the Kimberley in country that has been transformed into a white man's cattle station. The narrative reveals the lives of the Aboriginal people who struggled to survive under the imposition of slavery in the early twentieth century.

Aboriginal Traditions and Eco-criticism

Approaches to literary works based within eco-criticism foreground the elements of place and environment beyond what is manmade. Often this split between place and people follows Euro-centric traditions that oppose nature and culture, where the environment is a context rather than an active participant. As eco-critics have recognised, our environmental relationships are contingent on our cultural constructions of the self, the social and the physical (Moraga and Anzaldúa 23). The cultural constructs as well as the geographic location set up specific oppositions and understandings as well as contexts. As Kate Rigby and Axel Goodbody argue: "if there is such a thing as European eco-criticism" it would be geographically located within culturally and economically shaped landscapes given the long term and dense human impact on the space focusing on the pastoral rather than some ancient wilderness, in fact a domesticated or artificial space (3-4). As well as acknowledging that notions of the environment are geographically and culturally contingent, eco-criticism as an approach is applied to multiple contexts and conditions as well as to specific writers' sensibilities and localities. A problematic feature within current scholarship focused on minority literatures such as Aboriginal Australian is that they are often treated as a "state of exception" rather than another cultural formation that brings a specific perspective to the table. Because it takes as its subject every cultural position, eco-criticism potentially offers a way to extend current understandings without reproducing centre/margin paradigms.

According to Cheryll Glotfelty: "Simply defined, ecocriticism is the study of the relationship between literature and the physical environment" (Glotfelty "Introduction" xviii). This breadth of application beyond marginalised writers and groups offers a framework that can counter reductions to type and examine writers' work in a richer way. Aboriginal Australian writing offers a culturally specific contribution from a less Euro-centric perspective that actively engages with the physical world. Lawrence Buell defines work that is environmentally oriented through the following criteria:

- The non-human environment is present not merely as a framing device but as a presence that begins to suggest that human history is implicated in natural history.
- The human interest is not understood to be the only legitimate interest.
- Human accountability to the environment is part of the text's ethical orientation.
- Some sense of the environment as a process rather than as a constant or a given is at least implicit in the text. (7-8)

Drawing on traditional cultural understandings and connections, most Aboriginal writers not only meet these criteria but extend them. Aboriginal cultural perspectives start from a position where the land is a living presence that pervades everything. These cultural constructs establish a particular approach and understanding of the relationship between nature and culture that reveal the individual emotional and psychic toll of the disconnection from and abuse of the land.

A "fundamental premise of ecocriticism is that human culture is connected to the physical world" (Glotfelty and Fromm 14). Aboriginal traditions go further than this. Rather than being in an opposition between nature and culture, they are intrinsically interwoven within Aboriginal cultures. Seminal anthropologist W. H. Stanner sought to explain the incommensurability between Aboriginal and European understandings of land when he stated that:

> No English words are good enough to give a sense of the links between an Aboriginal group and its homeland. Our word 'home', as warm and suggestive though it be, does not match the Aboriginal word that may mean 'camp', 'hearth', 'country', 'everlasting home', 'totem place', 'life source', 'spirit centre' and much else all in one. Our word 'land' is too spare and meagre. We can now scarcely use it except with economic overtones unless we happen to be poets. The Aboriginal would speak of earth and use it in a rich symbolic way to mean his 'shoulder' or his 'side'. I have seen an Aboriginal embrace the earth he walked on. To put our words 'home' and 'land' together into 'homeland' is a little better but not much. A different tradition leaves us tongueless and earless towards this other world of meaning and significance. (230)

Traditionally, each Aboriginal grouping lived within their homelands or country that, according to their spiritual beliefs, was created and granted to them by their creator spirits. This territory is then passed down to each generation with the responsibility to know and care for the land. The physical world in all its aspects is part of a sacred bequest from the dreaming creators. This is not a relationship between living people and a passive physical world. Rather, it is a relationship across a continuum of life and the sacred.

This relationship to flora, fauna and the land is intrinsic to all social structures. Traditional Aboriginal society is structured around complex family or kinship systems. There are two main systems, kin groups and moieties. All aspects of life are organised around these systems and they locate an individual in society and in their country. These systems, kin groups and moieties, jointly help to determine many aspects of social or religious behaviour. Kinship is of fundamental importance in Aboriginal society. Aboriginal people regard themselves as being related (although not biologically or by marriage) to all the people within their cultural or linguistic region. In traditional Aboriginal society, everybody is kin and this method of reckoning can be extended to the entire social universe of Aboriginal people and beyond to non-Aboriginal people and all animate creation.

Individuals are further classified by moieties, a form of social organisation that divides people and natural phenomena into two classes or categories. Moieties bring everything together. The world of animate and inanimate forms is effectively divided into two halves: everything animate and inanimate, all people, flora, fauna and features of the natural and spiritual world belong to either one or the other of two moieties and are part of reciprocal relationships.

Moieties are usually named and associated with emblems or totems. For the Gunditjamara people in Victoria, the moieties are the krokitch, the white cockatoo and kaputch, the black cockatoo (Bourke and Edwards 110). The moiety works on ancestral relationships between people, animals and physical phenomena. For example, in the Kimberley, people belonging to the *Wodoi* moiety call the Spotted Night Jar their father. But they will also have other animal or plant associates.

To unproblematically link traditional practices and beliefs to contemporary writers risks the imposition of a nostalgic Aboriginalism. However, at the same time and to varying degrees, contemporary writers, often through family connections as well as anthropological knowledge, embrace and value traditional beliefs and viewpoints. Some writers have needed to discover through research and investigation the knowledges that have been lost to them as individuals, for others threads and layers of traditional values and world views have been absorbed through family stories, and for yet others the traditions continue to have currency in their daily lives handed down from their elders. Resulting from these different engagements with traditions, as each writer's sensibility is informed and shaped by his/her life experience and context, there are different points of emphasis in the writings of people with different cultural connections.

Regardless of the impact of colonisation on individuals' relationship to their heritage, land as country with emotional and spiritual significance is critically important to all Aboriginal writers in terms of their identity, their relationship with the world and their community. This is not in terms of a mystical connection but rather in terms of knowledge. The premises of Aboriginal teachings through song lines and creation stories were learning about the land, its needs and its conditions and how to survive in that specific landscape. Given the challenging nature of Australian conditions, this knowledge was, and in some parts of the country still is, essential. The Yanyuwa, an Aboriginal group in the Northern Territory, tell stories about tourists who are found starving and sometimes dying of thirst when they are surrounded by food and water that they cannot recognise. Several famous examples date back to the Second World War (Casey and Bradley 2011). Aboriginal traditions and stories are about the survival of both community and culture in relation to country as well as care of country.

Aboriginal Dramas in their Environment

Each Aboriginal writer's work reflects their specific heritage with all the complications imposed by colonisation. Some dramas are located in a particular country. Others present a more generalised perspective. Some writers such as Wesley Enoch alternate between the two for artistic ends. *Cookie's Table* is clearly located on Stradbroke Island, Enoch's own home country. Stradbroke Island is in Moreton Bay off the coast of Queensland. Stradbroke is a sand island with a heavy rainfall and substantial fresh water and dense bush. The Aboriginal name for the island is Minjerribah. Originally one island, it has been split into North and South Stradbroke by the Euro-Australian made Jumpinpin Channel. The Aboriginal peoples on Stradbroke are the Noonuccal, Goenpul and Ngugi peoples. Enoch is primarily Noonuccal. The Aboriginal people of Quandamooka still maintain their heritage and identity, treasuring the stories of their people and ancestors. There are story keepers who seek to collect and pass on tales of recent and distant forebears.

The narrative of *Cookie's Table* is informed and shaped by aspects of contemporary life on the island. From the dependence on the barges and water taxis to reach the mainland, through the impact of storms, to the tight family communities that continue to claim space and a strong cultural presence on the island despite all the complications of systemic racism, poverty, and the resulting alcoholism and sexual and domestic abuse.

In the Writer's Note prefacing the text of *Cookie's Table*, Enoch describes the play as "a metaphor for culture lived, lost, found, obscured and metamorphosed into a hybrid reality for contemporary Murri people" (n. pag.). The text illustrates the "unbroken line of cultural practices, through the feeding and gathering of our families around the table" (n. pag.). Enoch goes on to argue that, in the context of Native title claims to land, this community connection and survival is as meaningful as anthropological evidence about middens and camp sites. This constitutes a reference to the terms of connection demanded by courts under the Native Title Act 1993 that was amended in 1998. The Native Title Act is the basis for any recognition of Aboriginal land rights under Australian law. Anthropological evidence is primary. A famous example of the problems Enoch is referring to is the Yorta Yorta case for recognition of their land rights that was dismissed in 1998 and then refused after appeal to the High Court in 2002 on the basis that "the tide of history had washed away any real connection to tradition and the land" (Federal Court). The court rulings depended on a view of real Aboriginal culture as frozen in a pre-contact time and contemporary family stories and connections were ruled irrelevant.[1] To Enoch, *Cookie's Table* is a story about the strength of family. For him this continuity of family is primary evidence for the connection to ancestors and country. The text reveals many layers of this connection through family to country and the physical world and traces of its important place within Aboriginal traditional practices.

The table of the title and a central character in the play is all that remains from Cookie's birth tree. In traditional Aboriginal practices, birthing trees were an important part of physical and spiritual life. The moment when a woman went into labour, she was taken by other women to a birthing tree. There a shallow hole would be dug and lined with the bark of the tree to receive the child as it came out of the womb. In this way, the child was literally born into the land. The birth would be accompanied by smoking and song ceremonies. The birthing trees, as Nathan recounts in the play, were specially chosen by a mother or grandmother because:

> We believe the tree you are born under gives you something… like part of its spirit—and this tree was chosen 'cause it was strong with a wide trunk, yet supple enough to bend in the wind to survive fierce storms… While [… Cookie's] mother held onto a branch surrounded by her aunts [… she] was born. (Enoch 1)

As the opening monologue continues, we learn that after the white men came, "they cut down her birth tree [and] she felt it… like something had been torn out of her. She was in a canoe out on the bay fishing, and she felt like something had hooked her in the chest, was pulling her in, reeling her back to shore" (1). In the family stories, Cookie went to the place where her tree had been and found only a stump. She then followed the loggers' path until she found where the tree was and tracked what they had done to it. They made the lumber from the tree into a kitchen table for

[1] For further discussion of the Yorta Yorta case see Aileen Moreton-Robinson (2004).

white settlers. Following her birth tree, Cookie sought employment in the kitchen where the table made from it was sent (cf. 1-2).

The birthing ceremony is an important link to the land to which an individual has spiritual connection. The connection to the island that was her home is a strong thread in Cookie's story. It is transparently located on Stradbroke Island, through the connection to the ocean, and the references to the bay, the *dugong* and the turtle (1). It is this connection and the choice of tree that gives Cookie her gifts and strengths in cooking and survival. She stays strong because she maintains her immediate connection to her "beginning place, her birth tree" (2). The distress and dislocation caused by the denial of access to birthing trees is referred to in other works by Aboriginal playwrights. Ningali Lawford in her autobiographical one woman show, *Ningali* (1996), speaks of her grandmother's grief because the settlers had fenced off her birth tree and denied her access. She could only see it in the distance through a fence.

The main action of *Cookie's Table* focuses on the struggles for survival and dignity in the face of systemic racism affecting Cookie's great granddaughter and great great grandchildren. Throughout the arguments and tensions between three generations, Faith, Cookie's granddaughter, her daughter Annie and Annie's son Nathan, the table brings the focus back to place, kin and identity. After Faith's death, Annie and Nathan meet after the funeral and fight for possession of the table. Throughout the play, stories of Cookie and the table are interwoven within scenes and as separate stories that mark the narrative. These range from benign stories of Cookie adapting tradition and giving birth to her own children on the table (7) to the trauma of the fire that took Cookie's life and her sons.

After Cookie was deserted by her white partner, her rage became so great that she burned with it. Cookie was so full of anger that she made the white people she cooked for and served food to become sick. She was burning with a fever of anger so fierce that, according to local stories, she was thought to have finally burst literally into flames and caused the conflagration that consumed her, her family and her home. This fire can be understood as both literal and allegorical. For Enoch, the text is a metaphor for culture. As Aboriginal culture is intrinsically linked to the living land, it is not surprising that the metaphor is as much about the land as it is about the people's culture. Cookie is consumed by her fiery rage because she has been used and deserted by the white father of her children. This in turn makes her an active part of the land that overwhelms those that abuse it and continues even after they are dust. Fire is an unavoidable part of the Australian landscape and life. Fire has been used by Aboriginal people for millennia to landscape the country for hunting and the maintenance of plants. There are many Australian plants whose seeds only germinate after being exposed to fire. At the same time, fire can overwhelm those who do not respect the power of the land and climate. Rather than living in fear of fire, Cookie becomes fire. Her use and abuse by white people becomes a symbol of the treatment of Aboriginal people. She becomes an allegorical warning to those who would abuse the land and its people.

In addition, Cookie's fire affirms the continuation of Aboriginal Australian presence and cultures. When the European house was burnt to nothing, "the only thing left was this table standing in the middle of the ashes" (14). The image of the survival of the table as the only thing left can be understood in a range of ways, one of which is the survival of connection to land and, through that connection, to kin as part of the land. The table matters and lives because it is part of the birthing tree and therefore the land. Its place in the family is as a link to the land and the people. So, for Cookie and her descendants, each birth is carved into the table. The carvings signify the ways in which the people are part of the land and mark the land. Cookie's daughter Kawana is the only member of the family to survive the fire. She lives because her mother sent her to stay with another Aboriginal family. After the fire, members of the Aboriginal community retrieved the table and gave it to Kawana, saying: "This was your mother's birth tree, that's why she left it, why the fire couldn't take it away. It's yours now" (14). In her response, "Let this table be a home for me" (14), Kawana is embracing the land as well as kin and culture. Neither conquest by colonisers nor fire could destroy what the table is and represents.

Throughout the text, the presence of country, the island and the sea embed the family, their cultural identity, their history and their struggles within a specific land. So too, the tree and names carved into it are a direct link with land, family and history. Through this link, the present emplacement of Aboriginal people in a colonised context is given expression. It also enables personal reconciliation with the reality of being an alienated racialised minority in their own country at the same time as it affirms that they belong. In the final sequence of the play, Nathan and Annie talk about burying the table so that a new tree can grow over it and everyone's names can be carved in it, connecting the people and the earth. The final words of the play express Nathan's desire to have carved on his grave stone: "He belonged here" (54). For Nathan the claim that he belongs in and as part of this land is crucial. The hurt and estrangement of Nathan's and Annie's lives are eased by reconnection with the land.

Whereas in *Cookie's Table*, Enoch focused on the specific connections to family and land in his own homelands, in *Black Medea* he shifts to a more general view of Aboriginality and home lands. *Black Medea* is a piece of theatre in which the power of the earth drives the characters to honour their heritage in spite of colonialism. Enoch's *Black Medea,* first staged as part of the Sydney Theatre Company's 2000 *Blueprints Season*, was subsequently reworked and produced in 2005 at Belvoir Street and the Malthouse. In *Black Medea*, Enoch uses Euripides' *Medea* to "find a depth of expression and spirit" (59) in order to examine the dynamics of racism and colonisation and its impact on Indigenous Australians.

In *Black Medea*, Enoch locates his adaptation of the Greek classic Euripides's *Medea* within the desert lands of Central Australia. Since the mid twentieth century, these lands have been understood as one of the iconic homelands of continuing Aboriginal cultures and traditions because of the highly publicised land rights struggles. In Enoch's text, Medea, a young Aboriginal woman with strong links to the land and traditions, falls in love with an Aboriginal man working for a mining corporation. After leaving her home and family for Jason, she is haunted by the spirit of her land and responsibilities particularly in relation to the links with land that should be passed onto her son. The spirit of the land and Law is personified as an older woman from her clan, who demands that Medea return to her cultural roots. This woman, as chorus and story teller, is an ambiguous figure, part ancestral mother, part family elder and in part the spirit of the land and duties that Medea has rejected on behalf of herself and her children. Medea's marital and social problems are all attributed by this spirit to the separation from cultural life and land.

In her opening battle cry, Medea pits her knowledge as a daughter of the land against the spirit of the land that haunts both her and Jason. Within this battle is the relationship to land. People of the coast and islands are sea people. People of the desert are not. The haunting spirits and Medea's heart contrast her land that is the open desert with the coastal cities where "strangers [clutch] to the coast like cowering children" (63). The first scene reveals that because of Jason and Medea's choices to separate themselves from their heritage, the wind from the desert frays their nerves, exacerbating violence between them (64). In a discussion of Gernot Bohme's ecological aesthetics, Kate Rigby translates Hermann Schmitz's analysis of experiences as the result of "unlocalised, poured forth atmospheres … which visit (haunt) the body which receives them … affectively…" (274). This haunting finds an embodied expression through the emotional response it provokes. In Aboriginal traditions and contemporary writings, these atmospheres are localised and emanate from specific sources, particularly the land and the spirits of the land. As the winds and storms of Stradbroke affect the moods and feelings of the characters in *Cookie's Table* on subtle levels, in *Black Medea* the wind and the atmospheres it creates are a potent force that batters and overwhelms the characters. This is an atmosphere complete with moral and philosophical power with active intentions.

For Medea, the land that is part of her has become a threat. When she is with Jason, she is in fear of the wind "singing" him away (63). Yet, when she dreams, it is of the connection between herself and the land. When Medea recounts her dream, she tells of "staring into the desert" and knowing that she would never be alone (66). She sees generations of her maternal forebears, "an ocean of women stretching back out into the desert, stretching out to the horizon making the sand dark" (67). Strengthening the way in which these mothers and grandmothers are part of the land is the description of her grandmother as a "woman of law and language, standing ankle deep in the sand" (66). This symbolism of connection to land resonates with ceremonial dances. Traditional dance styles are specific to the type of country in which they are performed. The physical environment, the geographical location and conditions, are reflected in the types of physical movements within Aboriginal dances. Indeed, a dance from the desert will include sliding and shuffling movements representing the sand and wind. The physical movements express the connection between the earth and people and also the life and movement of the land (Casey 27).

Though there is some suggestion that Jason's people were originally from a coastal region or an island (Enoch 70), he has "no dance, no song, his uncles have never walked him through his country" (68). This dislocation and disconnection from his home lands and knowledge is the legacy or rather lack of legacy he has to pass on to his children. Jason's career in the mining industry adds another layer in the engagement with the environment in *Black Medea*. Mining as resource stripping without returning anything to the land adds to the lack of connection to land and traditional Aboriginal relationships with the land. Jason feels no responsibility to care for country. The land is a resource to be exploited until it is empty. The approach is disconnected from traditional ways, seeking profit without care for the future or the past. Medea not only walks away from her family and duties, she betrays her connection and knowledge of the land by revealing to Jason sacred places that offered him "the largest vein of that which" he would mine (74-75). Even Jason initially baulks at this, "for what sane man would rape his mother?" (75). But in the service of wealth, he agrees.

The issue of mining is a major one in relation to Aboriginal lands. It is a source of conflict within communities as the pressures of poverty and the potential of even inadequate mining royalties war with the protection of sacred sites and custodial responsibilities and connection with country. This betrayal of the knowledges Medea was privileged to have stands in conflict with her sense of identity and belonging that stems from her connection to country and traditional knowledge. Even after deserting the "dances and songs" that were hers, she deals with the world from the position of connection (72). Her very physical being expresses her connection to country. As she states:

> When I go down the street people stare at me. Whitefellas, blackfellas. They know I'm from somewhere they've never been… and they're scared. I remind them of what they don't have. They walk around like they are scared to put a foot wrong, apologising with every step—too scared to admit it, admit they don't belong here. I've known the spirits to come up through my feet and take my body when I'm dancing. (73)

Medea's connection to the land is inescapable because of her responsibility to the next generation. In the end, the land reclaims her, as she feels she must release her son back to the spirits of the land. As the Chorus says:

> You can't really leave your land
>
> She [Medea] reckons she left it behind with her family, left it behind like a footprint and every time she walks on foreign soil that foot speaks to her telling her it has a memory of where it belonged. You're related to the sand dunes and the waterholes. You belong in your land. The Land will always reclaim you. (78)

This active and expressive aspect of the land and the spirits embedded in it recurs in other works, such as David Milroy's *Windmill Baby.*[2] This play is a monologue for an older woman call May May. She grew up on a cattle station in the first half of the twentieth century when Aboriginal workers were basically bound as slaves working for rations, their lives under the control of and at the mercy of the station owners. May May has returned in the present to grieve and finally lay to rest people both Aboriginal and non-Aboriginal, including those of two newborn babies, whose lives were cut short by the racialised context that devalued their lives. The world she knew was one of hard labour and violence. In the present, helicopters and machinery do the work that once ruled the lives of her family and community.

In May May's past, the Aboriginal community that she was part of worked on the station and the white owner had complete control over their lives. He had little regard for the humanity of the Aboriginal people. However, the station owner's new wife attempted to deal in a different way with the Aboriginal people living as part of the station.

Reciprocity is a basic tenant of traditional Aboriginal life. In a world where the land and its gifts are not tame or easily accessible, this notion of constant exchange sustains life and communities as well as the land. Medea sought to escape the obligations of reciprocity when she denied her obligations to the land at the same time as she desired to profit from the knowledge of the land she had been taught. In *Windmill Baby*, this reciprocity is given physical form through notions and needs for balance. The first born of twins, Wunman, had "a crippled-up arm…

[2] David Milroy is a descendant of the Palyku people of the Pilbara (201)

and a crippled-up leg" (209). This left him precariously balanced in the world until he was given responsibility for the vegetable garden. His dead twin, Twoman, comes in the night to help him in the garden. As Wunman belongs to the land, the previously struggling garden responds to him and flourishes. Despite his lack of physical balance, he feels capable of balancing relationships between others, bridging racial barriers by inviting others such as the station owner's wife to belong in the land. A relationship based on reciprocity develops between Wunman and the station master's wife. Wunman gives her fruit and vegetables and she gives him music and an exchange of knowledge. From this reciprocal relationship, a more intimate one develops that results in both his death and that of the child from the relationship. At different points in the text, Wunman prevents the owner from killing other Aboriginal people but, in the end, he has no way to defend himself.

In *Windmill Baby*, the land and the spirits of the land are presences that watch, enclose and sometimes communicate with the people living within it. The spirits of the past are so embedded in the land that plant life moves as a gentle reminder of people and events. In the stage directions, a melon rolls across the stage that May May recognises as a message from Wunman (Milroy 209). This connection to animate and inanimate elements of the environment resonates with Aboriginal traditions that see all things in existence linked through moieties as kin. The name of the play comes from a spirit baby that is seen near the windmill, foreshadowing events that will happen. This resonates with the traditions of the spirits of the land giving human babies spiritual identity. Totems are the result of the mother being made aware of the spirit entering her when she is pregnant. The spirit comes to the unborn baby from the earth through the mother. That spirit is released again into the earth through traditional funeral rites. Though the apparent focus of the narrative is on the lives and deaths of people that May May knew in the past, her unfinished business is the need to acknowledge this continuing living spiritual essence in the land. Her acknowledgement of this presence allows balance to return.

Conclusion

In traditional Aboriginal cultures, birth and death as well as life are intrinsically linked to the living land. A striking theme of each of the plays discussed here is the human need to live in and with the land in a balanced and reciprocal relationship. In each text, the happiness and unhappiness of the human characters are related to the imbalances in connection to their relationship with their homelands. When the people and the land are in balance as a reciprocal relationship, individuals can produce what they need, cross racialised boundaries and know the potential for positive lives. When this relationship is out of balance, the atmosphere that surrounds people damages them. There is a human price to pay when the rules of reciprocity are broken. Medea's use of her traditional knowledge to feed Jason's desire for wealth, even more than renouncing her home lands, broke the pattern of reciprocity. Cookie's desertion by her white husband was a denial of reciprocity within their relationship and provoked the fire. Her family can only find peace and reconciliation when they reconnect with their home land. For May May, there could not be peace until those who had joined with the land through death were honoured through the recounting of their stories. In each of these plays, respect for the land on multiple levels is an essential element for happiness and survival.

The relationship of Aboriginal people with the country they are entrusted with by the creator spirits is based on mutual maintenance. The land must be cared for, respected and valued. This duty is to a living presence that incorporates Aboriginal past and future, not just a respect for physical objects separate from human life. As Stanner observed in the passage quoted earlier in this essay, traditionally Aboriginal people could literally embrace the land they walked on and valued it as a physical part of themselves. Though cultural experiences and beliefs have changed or been broken, the work of contemporary Aboriginal writers resonates with this deep connection to a living earth that is part of their emotional and physical lives.

Works Cited

Bourke, Colin and Bill Edwards. "Family and Kinship." Ed. Colin Bourke, Eleanor Bourke and Bill Edwards. *Aboriginal Australia* Second Edition. St Lucia, Queensland: U of Queensland P, 1998. 100-21.

Buell, Lawrence. *The Environmental Imagination: Thoreau, Nature Writing, and the Formation of American Culture*. Harvard: Harvard UP, 1995.

Casey, Maryrose. *Telling Stories: Aboriginal and Torres Strait Islander Performance*. Melbourne: Australian Scholarly Publishing, 2012.

Casey, Maryrose and John Bradley. "Aeroplane Dance: Whose Story is it?" *Screening the Past.* Issue 31 (2011). Web. 3 August 2013. http://www.screeningthepast.com/?p=1055.

Drake-Brockman, Henrietta. *Men Without Wives.* Sydney: Angus and Robertson, 1955.

Glotfelty, Cheryll. "Introduction." Ed. Cheryll Glotfelty and Harold Fromm. *The Ecocriticism Reader: Landmarks in Literary Ecology*. Athens and London: Georgia UP, 1996. xv-xxxvii.

— and Harold Fromm, eds. *The Ecocriticism Reader: Landmarks in Literary Ecology*. Athens and London: Georgia UP, 1996.

Goodbody, Axel, and Kate Rigby, eds. *Ecocritical Theory: New European Approaches.* Charlottesville: U of Virginia P, 2011.

Enoch, Wesley. *The Story of the Miracles at Cookie's Table*. Sydney: Currency, 2007.

—. *Black Medea* in *Contemporary Indigenous Plays.* Sydney: Currency, 2007. 55-81.

Federal Court of Australia. *Members of the Yorta Yorta Aboriginal Community v Victoria & Ors*. (1998) FCA 1606 (18 December 1998). 11 September 2011.

Lawford, Ningali and Angela Chaplin. '*Ningali.' Wimmin's Business.* Festival of the Dreaming. Opera House, Sydney, September, 23, 1997.

Milroy, David. *Windmill Baby* in *Contemporary Indigenous Plays.* Sydney: Currency, 2007. 201-27.

Moraga, Cherríe, and Gloria Anzaldúa. *This Bridge Called My Back: Writings by Radical Women of Color*. New York: Women of Color Press, 1986.

Moreton-Robinson, Aileen. "The Possessive Logic of Patriarchal White Sovereignty: The High Court and The Yorta Yorta Case." *Borderlands ejournal*. 3.2 (2004). Web. 15 April 2013. http://eprints.qut.edu.au.

Rigby, Kate. "Gernot Bohme's Ecological Aesthetics of Atmosphere." Unpublished paper.

Stanner, W E H. *White Man Got No Dreaming: Essays 1938-1973.* Canberra: Australian National University, 1979.

Dance, History and Country

An Uneasy Ecology in Australia

Rachael SWAIN

Co-artistic director, Marrugeku

An Uneasy Ecology[1]

The nature and function of Indigenous storytelling can be considered through a number of frames—mythic, historic and contemporary. That Indigenous myth narratives have a co-presence as land form, story, dance, song and what the Kunwinjku call *Djang*, or The Dreaming,[2] is widely understood by Indigenous cultural custodians and has been documented over decades of Australian ethnography.[3] *Djang* stories in their own "socio-topographic" context (Berndt & Berndt) function as both a mapping of country and a mapping of social interactions and in both the past and a continuous present as a living interpretation of what in non-Indigenous contexts we might call "existence." Stories are polysemic and devolve meaning in multiple ways simultaneously to different sections of the community. They are both universal and archetypal guides to multiple aspects of daily life, good and bad, but also operate as an enigmatic and somehow shifting veil, obscuring and revealing meaning in a complex and multifaceted patterning of narrations of events, transformations and kinship ties.

1 Warning: this essay contains references to and images of deceased people that may be sensitive to some Aboriginal people from communities in Western Arnhem Land and The Kimberley. Please consult with knowledgeable local people before sharing it with members of these communities.

2 The Dreaming or The Dreamtime are terms used by non-Indigenous Australians to attempt to name in English the Indigenous Australian concept of *songlines* both as creation stories and as a continuous living presence. I prefer to use the terms utilised by the two language groups I have worked most closely with. For the Kunwinjku the term is *Djang* and for the Yawuru it is *Bugarrigarra*.

3 See Deborah Bird Rose, Stephen Muecke, Marcia Langton, and Michael Jackson, to name a few.

Contemporary Indigenous storytelling and its relationship to country is significantly influenced by the social and political histories which have effected Indigenous knowledge transference and collective remembrance. This in turn informs the development of contemporary Indigenous art practice. In 1937 the first Commonwealth-State conference on "native welfare" adopted "assimilation" as national policy. "The destiny of the natives of aboriginal origin, but not the full blood, lies in their ultimate absorption…" (Human Rights and Equal Opportunities Commission 11). The children who were removed from their families under these acts are known as The Stolen Generations. The *National Inquiry into the Separation of Aboriginal and Torres Strait Islander Children from Their Families* concludes that "between one in three and one in ten Indigenous children were forcibly removed from their families and communities between 1910 and 1970" (Human Rights and Equal Opportunities Commission 4).

These genocidal policies resulted in the loss of connection to parenting, family, the practices of song, dance and language and the transmission of Indigenous knowledge systems and direct ceremonial connection to country. These policies had far-reaching consequences in many areas of Australia. The high number of forced removals resulted in the loss of many traditional songs, dances and stories and with these losses, connection to country took on very different meanings.

Together the complex and painful histories of these issues meant that the development of Indigenous performance practices drawing on a relationship to country and history is frequently steeped in loss, removal and at times massacre and can evoke the ghostly presence of multiple cultural traditions and stories. The cultural production of contemporary Indigenous art and its inherent culture mapping can require cultural revival and cultural intuition as a valid and important place to create from as an artist. Often the results in contemporary Indigenous visual and performing arts contain a potent mix of biography, fantasy and what I have termed cultural intuition.

Around the beginning of this century there was a collection of Indigenous and intercultural works developed *in situ* in remote Australia, each born out of a deep dialogue with Indigenous cultural custodians. I believe identifying these works as a small movement offers a bridge between the uneasy relationship to country of contemporary Indigenous arts and the deep and complex philosophy of Indigenous knowledge systems. I am proposing this gives the opportunity to analyze the possibilities of ecocritical Indigenous performance in Australia and its "uneasy" navigation of the interpenetrating aspects of mythic, historic and contemporary associations with country.

The Aboriginal English term country can denote an area of land formations and at times stories which a group or individual may have custodianship over. Indigenous leader Professor Mick Dodson explains that "[w]hen we talk about traditional 'Country'... we mean something beyond the dictionary definition of the word. For Aboriginal Australians... we might mean homeland, or tribal or clan area and we might mean more than just a place on the map. For us, Country is a word for all the values, places, resources, stories and cultural obligations associated with that area and its features. It describes the entirety of our ancestral domains" (Reconciliation.org.au).

By way of example, four works offering alternative approaches to Indigenous performance developed as a deep mapping of country include: *Trepang* created in 1999 by Yolngu performers from Galiwinku (Elcho Island) and Ujung Padang (Indonesia) directed by Andrish Saint-Clare; *Ochre and Dust* in 2000, an installation/performance created with visual artist Fiona Foley and Anangu-Pitjantiatjara story-women, Nura Ward and Nellie Patterson and directed by Aku Kadogo; Marrugeku's *Crying Baby* also in 2000, co-written by Kunwinjku storyman and painter Thompson Yulidjirri and myself and directed by myself and *Fire, Fire, Burning Bright* in 2002 by Peggy Patrick's Neminuwarlin Performance Group, performed by Giga and Mirrawong speaking artists of the East Kimberley and directed by Saint-Clare.

As opposed to most Indigenous Australian theatre and dance created in comfort and control in the city, each of these works were made in country, steeped in the micro signs of the physical environment, close to the scrutiny and guidance of cultural custodians of stories, songs and dance, whose authorship extended into the works through the complex net of kinship ties and responsibilities. Many of these works were also performed initially to a local audience and thus their dramaturgy was structured in response to local humour, perception of history and cultural knowledge.

In this essay, I will describe aspects of the "socio-topographic" contexts which surround the creation of contemporary Indigenous performance in a direct dialogue with custodians of country in an attempt to identify key signifiers of ecocritical performance developed in remote Australia and how this might function. I will do this through an examination of the use of historic time and mythic time and its relationship to country in Neminuwarlin's *Fire, Fire, Burning Bright* and in a case study of the narrative structure of Marrugeku's *Crying Baby* which I co-wrote and directed with Thompson Yulidjirri for Marrugeku.[4]

[4] For further information on Marrugeku see "*Burning Daylight*—Contemporary Indigenous dance, loss and cultural intuition"(Swain *The Arts and Ethics*), *Performance and Cosmopolitics* Gilbert and Lo (2010), and *Burning Daylight*, a documentary on the making of the production *Burning Daylight* in Broome, dir. Warwick Thornton.

Time, History and Country

Indigenous notions of time and conceptions of history underpin cause and effect in story structures and their place in a community's collective understanding of being in place and time. Patterns of occurrences interconnected with others, and intersections of mythic time and historic time can present a reality radically different from non-Indigenous understandings. Anthropologist Michael Jackson states: "Aboriginal people construe history as ever present, and ancestral land assumes for them the same vital force that self and soul have for us" (7).

That chronologies can follow another kind of logic emerged repeatedly during Marrugeku's research and rehearsals in Western Arnhem Land. A map of a certain geographic region drawn in the sand would contain a number of stories that became the dramaturgical structure of a production. Their co-presence across time in a region was foregrounded in importance to the linear chronology of events they summarised. This notion has been commented on by various anthropologists, notably by Deborah Bird Rose. Extending these ideas, Cultural Studies writer Stephen Muecke connects the notion of the creator beings as a living presence in the landscape with Indigenous understandings of a multi-temporal approach to reality and story:

> The Aranda (now spelt Arrente) understanding of the simultaneous presence of an Ancestor in many places is indicative of the absence of cumulative time and an understanding that presence can be held in place, enduring. Since country endures, it is always already there, place becomes the horizon of temporality against which human finitude finds a rhythm. Country can hold several moments simultaneously, just as an Ancestor may be present at many places simultaneously. So indigenous Australian time, while not metrical, was periodised nonetheless and enacted in rituals that brought out ancestral power in ever-tightening rhythms, from the slowest cycle of the ancient dreaming, to the cycle of generations, to the life of one custodian, to the seasonal ritual cycle to the song cycle itself and its repeated phrasings and stamping of feet into the earth. (16-17)

Thus Indigenous perceptions of time as "ever tightening rhythms" have a specific impact on the relationship to country we are considering and the logic of place and events which may be reflected in live performance practice.

In order to consider how history is conceived in Indigenous contexts and how this might impact on an Indigenous ecocritical performance mode, it is important to consider post-contact storytelling and the "irruptions into the dreaming" (Sansom) induced by contact with strangers. I am proposing that this is exemplified in the performance *Fire, Fire, Burning Bright*. Neminuwarlin Performance Group presented the conception of a historic event entering into mythic modes of Indigenous storytelling to audiences at both Perth International Arts Festival and Melbourne International Festival of Arts in 2002. The dramaturgy of this event was a radical departure from much Indigenous theatre previously seen in Australia though indicative of a small theatre "movement" of works developed in close collaboration with custodians of country, significant to this essay.

Fire, Fire, Burning Bright

The first act of *Fire, Fire* was a narration of the poisoning and murder of black seasonal workers early in the twentieth century by a white station owner as retribution for killing a bullock. Notably, the all-Indigenous cast played the white station workers painted up in 'white face' in their naturalistic style of "anti performance." The combination of this style with Indigenous modes of storytelling functioned as a metacritique of Western theatre techniques, speaking back simultaneously to the venue where it was staged (for example, The Victorian Arts Centre in the Melbourne season) and audiences' perceptions of theatre.

The second act related the exact same events but, in clear contrast to the first, this time the story was staged as a *Joonba*, a song and dance cycle. The Joonba, which oral histories tell was initiated after World War I, related the story of the spirits' journey after the massacre, from their burning corpses, up the local hills, across country of other skin groups and then to the coast near the Port George Mission. The form of the traditional dance style, where ceremonial decorated and painted wooden boards depicting song-lines are carried, incorporated the path of the spirits in the telling of this historic event. This mystical "flip side" to the deadpan narrative of the first half of the production was an extraordinary example of the elasticity of "culture" and a rare and enigmatic if somewhat mystifying experience for festival audiences.

In discussing how the *Joonba* came into being, Peggy Patrick, the cultural leader of Neminuwarlin, stated in an interview prior to the Melbourne Festival opening that, "[a]ccording to oral histories, the shadows of a group of men who had been poisoned and burnt gathered in a cave and looked back at their burning bodies and started to sing their story" (Crawford n.pag.). This concept is explained in a little more detail in her biography:

> *Fire, Fire, Burning Bright* was a stage performance telling the story of the Bedford Downs massacre. It incorporated a Joonba, a song and dance cycle, given to a clever man by the spirits of the people who died. The Joonba had been performed in secret by relations in the East Kimberley and never shown to Europeans. Patrick first saw this Joonba when she was a girl of about seven. In 2000 Patrick's brother the late Timmy Timms and his brother-in-law Paddy Bedford revealed the existence of the Joonba and 'woke up' the

> song, rehearsing with Patrick and other relations to stage a performance at the Telstra Art Award in Darwin that year. (Artspace n.pag.)

Director Andrish Saint Clare saw this performance at the Telstra Art Awards and discussed the possibility of a production with Patrick. Reviewer Jonathan Marshall noted the function of the resulting performance as a generous public act of mourning:

> Indigenous people watching the original dances would have been familiar with the broad history of the massacre: the dances are largely unconcerned with questions of cause and character. Rather, they were a densely coded form of spiritual mourning, the Kimberley people using their language and mythos to make sense of events such as the effect of white settlement on geographic mobility and economic relations. The production did away with the patronizing assumption that indigenous culture is 'ancient and unchanging.' (6)

In this instance, a historic event in country and its disturbances in the ghostly realm became contemporary performance in the form of *Joonba*, operating in mythic modes but nonetheless historical for it. The role of the *clever man*, an Aboriginal English term meaning a sorcerer, or the one who can speak to the spirit world, as a vehicle for the birth of a new song cycle in modern times is significant to the process of the story taking form. Marrugeku (now written *marrkidjpu*), the name given to the company by Kunwinjku elders when we initiated the project *Mimi*, is the Kunwinjku language plural of *clever man*. This was presumably given to us for the resemblance of our stilt-dancing acrobats to the Mimih spirits, though perhaps it is the performance itself that opens a door to the spirit world, or to other perceptions of time and place. That is, the company's creation of the ecocritical performance was taking the role of the clever man itself, channeling an event into contemporary form. This remains an informed guess on my behalf, as too many of the old people who guided our process of culture mapping have passed away—such is the nature of the work.

I am proposing that *Fire, Fire* and a cluster of other productions which had their genesis in country, guided closely by law men and women, offer us a way of reading the potential of what Indigenous ecocritical performance could grow to be in Australia if it was given the conditions to emerge. In discussing Indigenous perceptions of time Muecke states: "Time is not opening out into an unknown future, but it is intensifying onto the present moment down through concentric circles of forces and meanings. The song and dance happening now have their reference in the place, the dreaming, the kin group and so on" (17). In the case of the Gija and Mirrawong artists from Neminuwarlin, the song and the dance happening now also have their reference in the history of atrocities endured. A traumatic historic event becoming *Joonba*, through the vision of a clever man and then being translated into performance for both local and urban audiences, also offers insights into a way the artists processed trauma and loss in the form of storytelling.

In the catalogue to the exhibition *Blood on the Spinifex* which represented many of the Gija artists who participated in *Fire, Fire,* including Peggy Patrick, Indigenous academic and activist Marcia Langton observes a significant link between Gija performance practices and the play's structure. Of note is Langton's citation of Graeme Cornwell's explanation: "The Gurirr Gurirr ceremony took the form of a 'palga,' a narrative dance cycle, a vehicle by which both current and historical events and traditional spirit stories can be revealed in public… dancers carried, for each of the song lines, painted boards or other constructions" (13). This understanding of a narrative dance style, a style that carries embedded in its weft the present, the past and the spirit world explains the structure of Neminuwarlin's contemporary production and how a dance style builds a link between the multifaceted nature of a traditional performance form and a contemporary production exploring a historic event in country.

Here I would like to propose that this phenomenon shares a link with postmodern performance practices where juxtaposition of multiple narratives and "realities" that resonate across space and time and a work as a constructive principle, allowing for the possibility of multiple meanings to accumulate. In their chapter on devising postmodern art making strategies, performance researchers Heddon and Milling describe that "[i]n place of a singular meta-narrative there is contingency, partial and plural perspectives and an understanding of the rules that pertain to 'narrative construction,' of making truths" (205). This furthers my proposal that modes of contemporary performance practice offer alternative and resonant methods to makers of contemporary Indigenous performance which could further a deep engagement with Indigenous knowledge systems in live performance form. The current dominance of a British text-based drama model in Indigenous theatre may have limitations in some contexts.

It is the driving force behind the order of things that is of interest here. In Indigenous performance developed in relationship with custodians for country, this might stem from what Muecke proposes as the role of Indigenous Philosophy in "keeping things alive in their place." In the context of Peggy Patrick and her relatives, the Gija artists, there is a gravitas to the "order of things." It is an order steeped in country and experience, in particular, experiences of loss and trauma. In discussing the references in the Gija paintings in *Blood on the Spinifex*, Langton observes: "[t]he paintings refer to the ancestral beings whose existence remains poised in places throughout the country, and to the celebrated pathways of their travels that network the land. And, as well, the life histories of the artists and their forbears are memorialised in the well-worn tracks of their ancestors, so that the suffering and tragedy of their lives can be endured" (12). Following the example of *Fire, Fire, Burning Bright*, ecocritical performance in remote Australia could perhaps be thought of, after both Langton and Muecke, as a mapping of multi-temporal presences 'poised and alive' in the 'place-event' of a live performance work.

Case Study: Dramaturgy as Deep Mapping in Marrugeku's *Crying Baby*

In order to identify how this mapping can function in an ecocritical performance mode in remote Australia, I will outline the genesis and narrative structure of Marrugeku's production *Crying Baby* in the following case study.

Marrugeku is an intercultural company creating contemporary dance theatre in remote Indigenous communities. Marrugeku was based in the land of the Kunwinjku people of Kunbarllanjnja (also spelt Gunbalanya and previously known as Oenpelli), Western Arnhem Land in the Northern Territory from 1995 till 2003 and has been based in the land of the Yawuru people in Broome in the north of Western Australia since 2003.

The company creates its productions in direct dialogue with custodians of country who participate in the research, set core concepts, view rehearsals and give permission for use of song, dance and story in the company's experimental contemporary dance process. The social, political, historic and topographic contexts surrounding the work have a direct impact on the "culture mapping" required to create the large-scale productions in each location.

In Marrugeku's second production, *Crying Baby*, painter and storyman Thompson Yulidjirri redraws a map of his country each night in the sand during the performance as he tells the interwoven stories of the *Marladj Djang*, or Orphan Dreaming, of the removal of his family from country by Reverend James Watson, the first Missionary into Arnhem Land and, finally of Yinginar, the Kunwinjku creation mother who walked across the country from the sunrise to the sunset placing all the first people in the land as babies. He marks them out spatially for the audience, ever renewing the mutually arising inter-subjective identity of the work and its relationship to country.

In a statement he made for the *Crying Baby* program notes in August 2000, Thompson summarised:

> First thing I can say, I think its *gamak* (very good) that new story and old story go together. I will start with Crying Baby, this place, somewhere behind Coopers creek. The story starts at Croker Island, finishes up at Gapari. That little boy he was crying, all the time crying, all the time crying for food. If you let that little boy crying all the time, little boy will make rainbow come, will kill you all. Like this mob, white man came and take children away, might be I can't see them

> when they grow, they go away from mother and father, no one looking after them. White man collect all them children. Take em away, never come back. We have to look after them kid, then that rainbow angry, coming to kill all. (1)

This dramaturgy of layerings of place and time clearly demanded us to develop a narrative structure for the production which functioned in multiple frames but was grounded in a specific location, involving a large geographic area in which several company members lived, and throughout which their stories travelled. We were required to develop a multivocal form of storytelling that shifted Marrugeku's dance-theatre practice into an interdisciplinary form. To this end we extended our collaborations to include the video art of Indigenous film-maker Warwick Thornton. Warwick's astute visual interpretations of the stories themselves in a multi-screened video installation and his creative representation on film of Kunwinjku notions of place, time and culture added the possibility of further patternings and associations to represent multiple realities in one moment.

Whilst researching the production in 1998, the company drove in three troop carriers overland for five hours to visit the site of the *Marladj Djang* (orphan dreaming), escorted by a tractor from the local outstation for the final hour.

Image 1. Thompson Yulidjirri seated near the Crying Baby Rock, of which he is custodian, Arnhem Land, 1998. Photography Rachael Swain.

Not long after this trip the constellation of stories that made up *Crying Baby* crystallised in the following process:

> One day while Thompson and I were working on the stories, which would inform the new work, I was trying to clarify with him details of a particular *djang*. As always these details were geographic as well as narrative. We took out a piece of paper and he began drawing a map on which he marked the elements of the stories. It started on a piece of A4 paper which quickly spread to another one and another, and then we got a big piece of cardboard and then we stopped. We got three pieces of cardboard and started again—it went on and on. By the time we finished about three days later, we had a document which included: a number of *djang* stories, a long history of Thompson's father being taken by 'Mr Watson'—the 1st missionary in Arnhem Land from Gumadeer to Goulburn, as well as a number of contemporary stories of the community and Marrugeku member's own experiences in the region. This map became the score for the multi-narrative performance, grounded in one geographic region, and our mutual experiences of that place. The map itself became the framework around which our varying understandings of performance could be processed in a multi-temporal dramaturgical structure, drawing firmly on lived experience of place. (Swain 1)

This actual map and the process of performance making which stemmed from it are clear examples of dramaturgy as deep mapping, and each of the points outlined earlier can be recognised in the resulting production, which I will outline in detail here.

Image 2. The cliff face showing the two tribes Ngaladj and Wurdbu turned to stone, Arnhem Land, 1998. Photography Rachael Swain.

In devising the work with Thompson, directly from the map of his country we drew together early in the process, we focused the stories into a prologue and three distinct story regions. The opening scene of *Crying Baby* is a dance theatre reflection on life in the Kunbarllanjnja community now. People sit around campfires or TV sets in family or "skin" groupings; others wander, accepting or avoiding eye contact as protocol demands. Fragments of traditional and contemporary dance are mixed with the rough and tumble of street fights, acrobatic moves and bursts of hip-hop.

The set itself reflects the co-presence of multiple stories of one region across time with the natural environment of pandanus trees and mangrove roots growing out of and through the built environment of an abandoned satellite dish, a phone booth and the wreck of a boat. Telegraph poles provide the rigging for aerial scenes and white sand reminiscent of Arnhem Land riverbeds covers the floor.

Thompson, together with the production's composer Mathew Fargher, introduces the three strands of the work, which Thompson draws and redraws on the sand during the performance. A live camera records his mapping and this is projected onto the satellite dish and the TV monitors as he narrates.

Throughout the piece, after each of Thompson's narrations, storytelling continues in stilt choreography, traditional and contemporary dance and Warwick's video art. As Thompson finishes the prologue, the Crying Baby itself appears suspended in an aerial act behind him. He continues narrating the Crying Baby story:

> This little orphan boy was crying all the time, day and night for food. He was hungry and nobody would feed him. At the same time there were two tribes Ngaladj coming from inland and Wurdbu coming from the coast, both travelling to Gapari for ceremony. The orphan boy had an elder brother, who went out hunting for goanna with his two dogs. When he came back after two or three weeks, the little boy was still crying all day and all night. The older brother could talk to the spirit world and because no one looked after the crying boy he made the Rainbow serpent come. The Rainbow serpent, Kunjikuime, was angry at the noise of his crying and angry that no one looked after that little boy. She made a big storm and flood and killed all the tribes people. They were all turned to stone and became the dreaming place Gapari. (Marrugeku)

As Thompson leaves the stage, six stilt walkers, costumed as the two tribes, Ngaladj and Wurdbu, flood the stage from alternate directions. In stilt dance choreography and aerial acrobatics, informed by Kunwinjku dance they perform excerpts of the Crying Baby story.

During the following narration the Kunwinjku dancers and "Mr Watson" begin the scene in a skeleton of a boat, as Thompson remembered it from his childhood on Goulburn Island.

> When Thompson's family heard of this Mr Watson camping at Gumadeer they came down to meet him. They went across with him by sailing boat to Goulburn Island. They landed at a place called Ambungidjba. It was near there that they built the first church and mission in Arnhem Land.
>
> When Watson asked, "Is there anyone else on the mainland we can bring to this place?" they said "yes" and they went to collect the tribes at Wadjbi, Wamarli, Aanuru, Ungarraba, Mainjinaj and Warlang. Some speaking Maung and others Gunbarlang, they all came with Mr Watson to the Goulburn Island mission. (Marrugeku)

Various scenes of Mr Watson unfold with a voice-over from archives of his diary notes. He rides a small bike through a surreal landscape, is chased by unseen, cheeky mimih spirits and gives a Christian burial to members of tribes who passed away from white man's illnesses.

Image 3. Sofia Gibson and Trevor Jamieson as Ngaladj, Perth 2001. Photography Jon Green.

The final story is performed as an epilogue. Thompson walks to the centre of the stage and sits down. One by one, the remainder of the performers and musicians join him and sit to listen. The scene is transparent and self-reflective; staged as if part of the process of the company's work, listening to Thompson in rehearsal.

> This last story is also the first; it's the story of Yingarna, the creation mother. Before everything; spirit, people and stories came from Yingarna, the creation mother. She walked across Arnhem Land from the sunset country in the West to the sunrise country in the East with a whole lot of dilly bags on her head and shoulders. Each woven bag was full of babies. These babies she placed across the land, leaving spirits; Nagiki, Namorrodo, Nammarnde, Yawk, Yawk and Mimi. All the tribes she placed on the land—the first people, and all their languages. She placed them bit by bit by bit, bit by bit 'till all her dilly bags were empty. (Marrugeku)

During the final rehearsals in Kunbarllanjnja in 2000, I was faced with the challenge of finding a closing scene for the work, one that had grown out of the process of mapping Thompson's country. How could one scene draw together the threads of each of the previous stories of displacement and dislocation, loss of family and country and function narratively yet remain anchored in the 'place-world' of the work?

I approached Thompson for his opinion on how to consider the final scene of the show. After thinking for a short period, he came back with the simple response that he wanted to finish the performance by telling the story of Yingarna, the Kunwinjku creation mother, and how she walked across the country and placed all the babies from her multitude of dilly bags into the land. The babies that became all the spirit people who populate the country as well as the first people, the first tribes. The elegiac poetics of finishing a piece exploring an Orphan Dreaming and his perception of its resonance with the history of the Stolen Generations, including his own family's experience of being taken from their country, with the creation story of Yingarna, is a testament to both Thompson as an artist and dramaturge, the radically different logic of chronologies I outlined earlier (after Neminuwarlin), a conceptual perspective of "seeing" country and our distinctive intercultural rehearsal process as deep mapping of a region of Aboriginal land in the north of Australia.

By way of an introduction, Thompson asked us to translate into English for the non-Kunwinjku audiences: "This last story is also the first," thereby giving the audience an entrée into the world of the Kunwinjku and everything as it began, when the babies were in the right place and the forces which forged the land and its people were in their fullest being. He offered this as contrast to the other stories of removal and abandonment contained in the production, as a simple story, told at the end of the work and as an epilogue.

In staging the scene, I asked all the performers to simply enter the space and sit down with Thompson, listening as he told the story. The video screens contained Warwick's contribution to the multiple layers of meaning in Thompson's choice of story. Warwick had set up a camera in the school grounds in Kunbarllanjnja, allowing the kids to take turns standing behind the camera to shoot footage of themselves for over an hour or so. The playful antics of the current generation of Kunwinjku children now filled the screens as a complement to Thompson's telling of the creation story.

Image 4. Dancers Eddie Nailibidj and Harry Thompson with Katia Molino as Rev Watson. Photography Jon Green 2001.

Belgian dance dramaturge Hildegard De Vuyst has stated, during a workshop with us on dance dramaturgy, that two of her most important functions as a dramaturge in devised dance are to recognise the 'inner logic' that a work is dictating as it evolves, and, once this is recognised, to propose ways for this to be 'radicalised' in the work (De Vuyst). Both these things were occurring in Thompson's decision to narrate the Yingarna story as the final scene. He recognised what De Vuyst calls the "dictate" the inner logic dictated by the work, and transformed this into an ending that counterpoints the loss of country with a story containing a deep sense of being in place. Thompson was functioning as a dramaturge in the work yet drawing on a deeply negotiated intercultural process and an Indigenous logic and perspective of "seeing" country.

Seeing Country

Jackson effectively discusses the relationship between the lived experience of place and the dreamtime:

> This relationship between lifetime and dreamtime is one of synecdoche. Past and present are brought together. Biography and mythology are effectively fused. What happens in the one is seen to have repercussions in the other. Thus the concerted effort of people in bringing the Dreaming into Being is reciprocally linked to the force that the Dreaming has in human conception and growth. (132-33)

This reciprocity emerged in the work with intermittent reminders that the rehearsal process was "bringing the Dreaming into Being" as expressed by Jackson above. Trips out into country would be seen to have triggered events in company members' lives, which in turn affected rehearsal processes and outcomes. Certain dances could not be danced in proximity to others, as I was told the dancers would get sick for dancing them out of context. Some songs could only be sung some of the time, for example, on alternate nights and at a certain place in the structure of the work, by song man, Bruce Nabegeyo or else he might get "weak." Another song when sung would always bring the wind.

Howard Morphy, who works at the nexus of art history and anthropology, explores visuality and representation in the bark paintings on the Yolngu of Eastern Arnhem Land. In an intercultural investigation, Morphy explores the shifting cultural and social contexts that surround the production of Aboriginal art, but also possibilities for European and Aboriginal conceptions of art to converge. He cites Richard Brettell: "Art is not so much a mode of representation as it is a mode of seeing, and that in making important representations artists present the viewer not simply with an internally consistent image, but with an entire world view" (87).

The notion that the world view, which is the result of the "way of seeing," applied by Indigenous artists to their practice can be extended to the cross-cultural performing arts and the representation of reality encapsulated in the dramaturgy of the work. Morphy notes that as hunter gatherers, the Yolngu develop an "acuity of vision" that seems remarkable to the outsider. Yet they do not represent the subtleties of the landscape or the water as they paint (89). "Yolngu do represent such features in art, and in music and dance, but they do not represent them as they are seen" (89).

It is the relationship between the seen and the unseen that is a critical element in Indigenous visual art. Here I propose this relationship between the seen and the unseen is also central to dramaturgy in ecocritical Indigenous performing arts. The perspective in Indigenous place-painting reveals an immersion in country, steeped in both its micro signs and its metanarratives which is relevant to a dramaturgy which evolves *in situ*, drawn from a perspective which is not "seen" but grown from sitting down (Ab. Eng.) in country.

Image 5. Yingarna story told by Thompson Yulidjirri and translated by Mathew Fargher. Photography Heidrun Lohr, Sydney 2001.

Conclusion

I have outlined an ecocritical performance mode which asks the spectator to pay attention to country and its stories by seeing, listening and feeling, an intercultural performance language that is incomplete, polysemic, knows of haunting and stirs ghosts. As a dramaturgy developed in the native title era it is a dramaturgy where dispossession/possession of place and belonging are interrogated without resolution in an uneasy ecology.

This ecocritical form of performance functions in poetic, multi-vocal and symphonic modes which I propose can achieve more complex and nuanced results for Indigenous storytelling than Indigenous theatre following a British text-based model with its conventional approach to narrative and drama.

If we are able to open spaces for new forms of cultural production to emerge from an engagement with Indigenous knowledge systems, country and interdisciplinary modes of performance we may come to be able listen to history, to sense and experience our environment and its stories. We may be able to better understand experiences of loss and disconnection from place and time. We may be able to reconnect the experience of art with social, political and cultural domains that shape our world.

Works Cited

Artspace, Raft. "Peggy Patrick—Dirrmingali, Barratjil—Naangari (C. 1930)." Web. August 2013. www.raftartspace.com.au/artists/PeggyPatrick.html.

Berndt, Ronald M., and Catherine H. Berndt. *The Speaking Land: Myth and Story in Aboriginal Australia*. Ringwood, Vic.: Penguin, 1989.

Crawford, Ashley. "From Muster to Grandeur for Festival Act." *The Age* 7 (Oct. 2002): n. pag.

De Vuyst, Hildegard. "Dance Dramaturgy." Sydney, 2010. Workshop.

Dodson, Mick. 2010. Welcome to and Acknowledgement of Country Factsheet. 24 March 2010. Web. November 2012. http://www.reconciliation.org.au/home/reconciliation-resources/facts-figures/qa-factsheets/welcome-to-and-acknowledgement-of-country.

Gilbert, Helen, and Jacqueline Lo. *Performance and Cosmopolitics: Cross-Cultural Transactions in Australasia*. Basingstoke (England); New York: Palgrave Macmillan, 2007.

Heddon, Deirdre, and Jane Milling. *Devising Performance: a Critical History*. Basingstoke (England); New York: Palgrave Macmillan, 2006.

Human Rights and Equal Opportunities Commission. 1997. *Bringing Them Home*: Guide to the Final Report. Canberra: Human Rights and Equal Opportunities Commission.

Jackson, Michael. *Minima Ethnographica: Intersubjectivity and the Anthropological Project*. Chicago and London: U of Chicago P, 1998.

Langton, Marcia. *Blood on the Spinifex*. Ed. Ian Potter Museum of Art, the University of Melbourne. Melbourne. 2002.

Marrugeku. *Crying Baby Program Notes*. Ed. Marrugeku, 2000.

— and Yulidjirri, Thompson. "Crying Baby." Dir. Rachael Swain. Perf. Thompson Yulidjirri. 2000. Performance.

Marshall, Jonathan. "Fire, Fire, Burning Bright." *Real Time* 52.1 (2002): 6.

Morphy, Howard. *Becoming Art: Exploring Cross-Cultural Categories*. Sydney: U of New South Wales P, 2008.

Muecke, Stephen. *Ancient & Modern: Time, Culture and Indigenous Philosophy*. Sydney: U of New South Wales P, 2004.

Neminuwarlin Performance Group. "Fire, Fire, Burning Bright (Marnem, Marnem, Dililib Benuwarrenji)" Dir. Andrish Saint-Clare. Perf. Peggy Patrick. Perth. 2002. Performance.

Nura Ward *et al*. "Ochre and Dust." Dir. Aku Kadogo. Perf. Nura Ward, Nellie Patterson. Perth. 2000. Performance.

Saint-Clare, Andrish. "Trepang." Dir Andrish Saint-Clare. Perf. Macassan Artists from Sulawesi and Yolngu artists from Elcho Island. Darwin. 1999. Performance.

Sansom, Basil. "Irruptions of the Dreamings in Post-Colonial Australia." *Oceania* 72.1 (2010): 1-33.

Swain, Rachael. "Contemporary Indigenous Dance, Loss and Cultural Intuition." Ed. Paul Macneill. *Ethics and the Arts*. Amsterdam: Springer, 2014 (forthcoming).

—. "Marrugeku's Site Specific Theatre Practice." Performance Space, Sydney, 2002. Conference Paper.

—. *Ways of Listening. Dramaturgy as Deep Mapping in Intercultural-Indigenous Performance*. Diss. University of Melbourne. Melbourne. 2010.

“Je te parle d’harmonie entre les plantes”

Ecologies of New Caledonian Nationhood in Pierre Gope’s *La Parenthèse*

Diana LOOSER

University of Queensland, Brisbane

Inside the garden we can construct a country of our own.
(Robert Louis Stevenson “The Ideal House” 200)

This essay considers *La Parenthèse* (2004) by New Caledonia’s premier Kanak (Indigenous) playwright, Pierre Wakaw Gope, in terms of its ecocritical engagement with the nation’s (post)colonial milieu. Shortlisted for the 2006 Grand Prix de littérature dramatique, Gope’s play uses botanical metaphors and plants as dramatic characters to pose challenging questions about New Caledonia’s history, current political temperament, and future directions. The garden setting, with its various endemic, imported, marginalized, and parasitic plants, structures an allegory for the nation and its various uneasy constituents in the wake of the violent civil uprisings between Kanak and French during the 1980s (*les Événements*). The plants’ growth, competition, destruction, and renewal during the course of the drama tender a trenchant appraisal of the ways New Caledonia’s populace and its key political and cultural institutions mediate competing tensions between Indigenous custom and Western innovation, and negotiate various modes of national belonging. Of particular concern to Gope is how the legacy of Kanak independence leader Jean-Marie Tjibaou has been maintained in the years since his assassination in 1989, especially the extent to which Tjibaou’s vision of a harmonious and mutually sustaining relationship between members of the nation-state, undergirded by a modern Kanak culture enriched by the past but looking to the future, has been supported or undermined by those who are its ostensible trustees. The natural environment forms a reflexive basis for the play, consistently reminding audiences of the deep, reciprocal, and fragile associations between land, culture, and identity. Yet Gope’s deployment of nature as an agent for interrogating national policy suggests how a focus on the ethics of environmental practices might engage a wider array of issues relating to cultural politics. Gope's play thus suggests the potential for an expansion of ecocriticism as practice, and foregrounds theatre as a potent forum for this process.

La Parenthèse (Parenthesis, Bracket) forms part of a broader genre of postcolonial Pacific Island performances that, during the past few decades, have evaluated the environmental fallout from colonization and globalization, reflected and participated in ecological debates, and experimented with sustainable production processes. This turn is not surprising, given that Oceania's regional imaginary has been based, in no small part, on shared geographic and environmental concerns: the experience of living on islands, the ocean as a common heritage and responsibility, the conservation of indigenous lifeways and resources, the impact of climate change on land/marine life and sea levels, and anxieties related to development and foreign enterprise. Artists have taken various dramaturgical approaches in registering and elaborating these concerns: *inter alia*, the haunting *Birds With Skymirrors* (2010) by Samoan/New Zealand dance company, MAU, laments the environmental degradation of islands and sea with a special focus on Kiribati; Hawaiian dramatist Victoria Nalani Kneubuhl's youth drama, *Ka Wai Ola* (*The Living Water*, 1998) highlights the exigencies of water resource management; Alani Apio's plays *Kāmau* (1993) and *Kāmau A'e* (1997) raise questions about place and identity in contemporary Hawai'i, dramatizing conflict between aggressive resort development and the maintenance of a traditional fishing area; and Gope's own *Le Dernier crépuscule* (*The Last Nightfall*, 2001) examines the impact of nickel mining on a tribal village. Community theatre in Oceania has frequently treated ecological issues, as seen in William Takaku's dystopian anti-logging play *Conference of the Birds* (1989) from Papua New Guinea; and in the wide variety of environmental plays by Wan Smolbag Theatre in Vanuatu, some of which have spawned ecoactivist initiatives such as the network of turtle monitors inspired by *Mi Wan Totel* (*I am a Turtle*, 1995). Tahitian (Mā'ohi) playwright Valérie Gobrait's *Le Partage de la terre* (*The Sharing of the Land*, 2001) emphasizes holistic ties between self, land, and heritage—sentiments also expressed in several plays by Māori playwrights Briar Grace-Smith and Hone Kouka. Related New Zealand works include Apirana Taylor's anti-nuclear solo performance *Message in a Bottle* (1994); and Lynda Chanwai-Earle's *Heat* (2010). Detailing the interactions of human and animal characters within Antarctica's delicate ecosystem, Chanwai-Earle's play was a ground-breaking example of "green theatre" as the first production to operate independently of the national power grid. *La Parenthèse* refracts several of these ecocritical concerns and strategies, while extending them in alternative dramatic and political directions.

Born in 1966 on the island of Maré in the Loyalty archipelago, and a member of one of the clans of the Pénélo tribe, Gope studied theatre arts in Côte d'Ivoire with the Kotéba theatre troupe under Souleymane Koly; in Vanuatu with Peter Walker, director of Wan Smolbag Theatre; and in Rennes, France, with Peter Brook. This interweaving of European, Pacific, and African influences has led Gope to develop a distinctive style of theatrical expression—impassioned, acerbic, and satirically comic—that syncretizes Western dramatic structures and local language, customary references, and rituals to explore the dialectic between tradition and modernity. Gope's provocative dramas are focused as much on intra-Kanak dynamics as on settler-indigene dichotomies, and exhort both Kanak and settler audiences to work together towards a productive common destiny (Bihan 21). He has (co-)authored over 15 plays since the early 1990s, produced mainly with his theatre company, Cebue (meaning "memory" in Nengone, his tribal dialect),[1] some of which have toured France and the Pacific; other projects include a creative collaboration with actors from Réunion Island in the Indian Ocean, and another collaborative work has been translated into Italian.[2] Gope, then, is one of Oceania's more prolific dramatists, yet in the predominantly English-speaking Pacific and the wider Anglophone world his oeuvre is not as well known, with only one play currently available in English translation.[3]

Gope's situation speaks to a perennial problem regarding the dissemination of Pacific works beyond colonially imposed boundaries. As Eric Waddell notes, intellectuals and artists from the French Pacific tend to orient their references, networks, and values toward metropolitan France and other nodes of the Francophone world, and are "marked by events, struggles, and debates taking place within that universe" (12) rather than in other parts of the English-speaking Pacific or the general Anglo-American realm. To a major extent this is an issue of language, as Robert Nicole avers: "Pacific literature in English is all presence, visibility, articulation, and therefore power, while Pacific literature in French is all absence, invisibility, and silence," with entrenched French colonialism meaning that "many years after independence in other countries of the region, francophone Islanders remain conveniently gagged and largely unheard by fellow Pacific Islanders, a potentially powerful and sympathetic ally" (265). Although in recent years there have been increasing dialogues and collaborations between Anglophone and Francophone Pacific writers, much creative work and accompanying

---

1 One of 28 Kanak dialects officially recognized in New Caledonia alongside French.

2 *Gli dei sono Ciechi* (2003), a translation of *Les Dieux sont Borgnes* (2002), co-authored with Caldoche playwright Nicolas Kurtovitch.

3 *The Last Nightfall* (2001), a translation of *Le dernier crépuscule* (2001).

criticism still tends to circulate locally (Keown 241-42). Kareva Mateata-Allain emphasizes the importance of the inter-island exchange of art and ideas as a means of overcoming the intellectual isolation caused by French colonialism, characterizing literature in translation as a "metaphorical *va'a* [voyaging canoe] that crosses imperial boundaries to connect Pacific peoples" (46). Consequently, by bringing Francophone texts into the ambit of Anglophone criticism, my discussion contributes to a broader project of building bridges across cultural, linguistic, and colonial borders through scholarly interpretation, facilitating at once a greater appreciation of shared struggles, concerns, and artistic endeavors in various Pacific communities, and a more variegated view of theatrical output in contemporary Oceania.

Gope's work is indicative of a new phase and style of artistic expression in New Caledonia that has emerged from a tumultuous period in the nation's socio-political history. This history is not only useful for contextualizing Gope as a playwright, but forms a self-conscious foundation for the play's ecocritical allegory. A country of 256,000 people with a 40% Kanak (indigenous Melanesian) and a 29% Caldoche (European) population, New Caledonia/Kanaky is a *sui generis* collectivity of France, currently undergoing a gradual transfer of power from the French state to the nation. Annexed by France in 1853, the archipelago's changing role as a sandalwood trading post, penal colony, free settler colony, and nickel mining hub encouraged increasingly tenacious colonial interests, with Kanak suffering land alienation, political disenfranchisement, and social and cultural marginalization. A Kanak independence movement began in the late 1960s and became particularly visible during the 1970s. The most prominent activist for whom cultural revitalization was a key objective was Jean-Marie Tjibaou (1936-1989), the driving force behind the Mélanésia 2000 Festival (1975) that was central in helping mobilize a Kanak cultural renaissance. From the outset, Tjibaou was committed to a vision of a shared nation based on a harmonious interethnic co-existence. Faced, however, with an intransigent French republic, political negotiation gave way to sectarian violence during the 1980s, euphemistically dubbed *les Événements* (the events), and marked by murders, hostage takings, and an influx of French troops. The Matignon Accords, signed in Paris in 1988 between delegates of the French state and the leaders of the rival political parties in New Caledonia (represented by Jacques Lafleur and Tjibaou), were an attempt to restore peace, create economic balance, and institute steps toward decolonization, with a referendum on independence to take place in 1998. Although the agreement helped appease the crisis, it remained a problematic decision for certain factions of the Kanak community who wanted immediate, unilateral independence in place of compromise, dialogue, and deferral (Waddell 184). During Tjibaou's visit to the island of Ouvéa in May 1989 to commemorate the deaths of 19 Kanak killed there the previous year, he and his deputy Yeiwéné Yeiwéné were shot dead by Djubelly Wéa, a zealous Kanak nationalist who felt betrayed by Tjibaou's part in the Accords.

During the period of Kanak and cross-cultural renewal that followed the Matignon Accords and the subsequent Nouméa Accord (1998) (Maclellan 10-11, Jouve n.pag.), the most tangible signifier of Tjibaou's legacy has been the Tjibaou Cultural Center opened in 1998. Spearheaded jointly by the Agence de Développement de la Culture Kanak (ADCK)—a post-Matignon initiative—and the French government, and designed by Italian architect Renzo Piano, the world-class structure was intended as a national and global symbol of contemporary Kanak identity as conceived by Tjibaou, focusing on creative artistic expression. Located on the former site of Mélanésia 2000 on the Tinu Peninsula near Noumea, the complex evokes a stylized Kanak village that integrates traditional custom while emphasizing the unfinished and constantly evolving aspect of culture. The buildings utilize ecological architectural principles, combining state-of-the-art technology with natural materials and an imbricated relationship with the surrounding environment, incorporating a series of mythological gardens that trace the story of the Kanak people from creation to rebirth (de Blas n.p.). Yet as Waddell observes, the Center has also been a site of controversy due to its "luxurious and monumental character" (195); its position near the "white" capital city; its global outlook at the expense of local liaisons and patronage; and its financing by the French state under the aegis of François Mitterrand's *grands projets présidentiels*, which raised the question of whether the construction of the Center actually served to extend the influence of France in the Pacific via the rubric of restoring the identity of its colonized subjects. These concerns have been relevant to the country *in toto*. The absence of a visionary leader to replace Tjibaou; fractured, varying views on independence among Kanak, Caldoche, and New Caledonia's other social demographics;[4] and an emphasis on rapid economic development designed to integrate Kanak into the contemporary global economy at the expense of sovereignty and at the risk of massive environmental damage, have bred anxieties regarding the extent to which Tjibaou's vision for Kanak self-determination and national community has been upheld by the Center and the broader political hierarchy, and to what extent it has been undermined, redirected, or abandoned (Waddell 195-200, 206).

---

4 It is worth noting that 1998 marked the inauguration of the Cultural Center, but not the referendum on independence initially promised by the Matignon Accords. The Nouméa Accord makes provision for a new referendum to take place some time between 2014 and 2019.

It is this fraught recent history and its multivalent implications that Gope maps in *La Parenthèse*, using the metaphor of the garden and its dialectical relationship with the adjacent House to anatomize, on one level, the Tjibaou Center as a metonymic exponent of Kanak culture, and on another, the nation as a whole. The key concerns expressed by the plant characters in the garden (the general population) and the human characters in the House (the political and cultural leaders who determine the destiny of the garden) involve the rationale and expression of Kanak cultural identity and the appropriate guiding principles for a national community. In both cases, the play extends two options: an Indigenous sensibility grounded in traditional custom and ancestral memory as a foundation for building a new future, versus a cultural outlook reduced to folklore and prefabrication and diffused by Western global cosmopolitanism; and the ideal of national concord envisaged by Tjibaou, versus a splintered society riven by petty grievances, intolerance, and lust for wealth and power. The play's characterization of contemporary New Caledonia is deeply ambivalent, symbolized by the Manager's decision to tear out the wise and ancient Gaiac tree from the garden, leaving an enormous hole that cannot be filled satisfactorily.

Gope has staunchly maintained that *La Parenthèse* is not designed as an attack on the Center or the people who work there (58).[5] While the play might suggest certain directions for the Center's cultural policy, Gope is invested in the more profound question of "c'est ce qui empêche le cœur de battre, ce qui le fait engraisser, étouffer, ce qui fait que le pays ne marche pas, ne fonctionne plus" ["what prevents the heart beating, what makes it fat, choking, so that the country does not work, no longer functions"; 58].[6] This approach is fueled by Gope's conviction that "le Vieux[7] est mort avec tout son rêve, et que ceux qui devaient continuer, porter la flamme, le font pour moi à contresens. J'ai l'impression qu'ils éteignent la flamme, la braise qui devait être protégée, qu'ils sont en train de l'enterrer" ["the Old Man is dead with all his dreams, and those who continue to carry the flame do it, for me, in the wrong way. I feel like they put out the flame, they are burying the embers that should be protected"; 58]. By staging this intervention, Gope's drama exemplifies contemporary New Caledonian theatre as "un véritable laboratoire culturel"

---

5 Page references are from the published version of the play and the interview with Gope that follows it (57-65). This published text differs in certain respects from the performed version, but I am using the published edition because it is the version most accessible to other readers. As a result, my analysis places a greater emphasis on the play as dramatic literature.

6 Unless otherwise indicated, all translations from French are my own.

7 Le Vieux (the Old Man), a familiar name for Jean-Marie Tjibaou used throughout the play.

["a genuine cultural laboratory"; Fenoglio 44], with the playwright drawing on a career enabled by artistic developments in the wake of the Matignon agreements, but reflecting critically on the terms and conditions of those developments.

In deploying the garden archetype to analyze the socio-spatial logic of the nation-state and the vicissitudes of cultural history, Gope is engaging a gardenist discourse marked by shifting interpretations from the biblical to the postmodern (not to mention the overdetermined depiction of the "South Seas" island as prelapsarian paradise). Gardens are already highly symbolic; they are, in John Dixon Hunt's words, "concentrated or perfected forms of place-making" (11) that "reflect, answer, even create certain human needs and concerns" (13). Like the multicultural nation-state, and like the theatre itself, "Garden enclosures both define their spaces and appeal across boundaries—by way of representation, imitation, and allusion—to a world dispersed elsewhere" (Hunt 29). The Tjibaou Center gardens, which comprise one of the play's representational frames and infuse its broader national allegory, perform a microcosmic mimesis of a cultural cosmology that explodes the spatio-temporal constructs of the modern nation to encompass the entire spiritual and material heritage of the Kanak people. This strategy reminds audiences of the holistic life-world that binds the human and non-human, the organic and inorganic, and the living and the dead, and posits the land as the ground of identity and the primary contention in the play. At the same time, the garden trope offers up a series of particularly apt metaphors for scrutinizing multiple social operations (cultivation, parasitism, poisoning, symbiosis, colonization, invasion, uprooting, biodiversity, grafting, co-habitation, flourishing), which help create a necessary critical distance for audience reflection and action. As Gope explains, "Si les hommes n'arrivent pas à entendre, à écouter, peut-être les plantes peuvent parler à leur tour. Il est écrit que si l'homme ne parle plus, la pierre parlera" ["If people are unable to hear, to listen, then maybe the plants can speak in their turn. It is written that if the man no longer speaks, the stone will speak"; 60].

Tjibaou's ideal of an inclusive and mutually sustainable society is advanced from the play's opening scene, which references Tjibaou's *levée de deuil* (lifting of mourning) ceremony conducted a year after his assassination. The Tribal Chorus' eulogy to Tjibaou affirms that "C'était ton voeu le plus cher de faire de cette terre ton jardin. Un jardin où toutes les fleurs ensemble soient liées, ensemble rassemblées, unies dans un même esprit, avec pour seul souci de répandre le parfum de l'amour" ["It was your greatest wish to make this land your garden. A garden where all the flowers might be linked together, gathered together, united in the same spirit, with the sole desire to spread the scent of love"; 10]. For the Gardener, who fights to maintain the garden in keeping with Tjibaou's ideology, this means a nation in which "chaque homme, chaque femme, chaque enfant ait sa place sur cette terre, quelle que soit la couleur de sa peau, quelle que soit son appartenance ethnique, politique ou autre" ["each man, each woman, each child has a place in this land, regardless of skin color, regardless of ethnic, political, or other affiliation"; 10]. It is an idea of community that is grounded in a vested relationship with the land, that embraces indigene, settler, and exile, and cuts across socio-economic lines to include the criminal underclass, the exploited, and the poor, scattered and marginalized by colonial modernity. As well as spreading trees and ornamental plants, the garden embraces "des fleurs sauvages poussées au gré des vents, qui cherchent un peu de lumière dans les fissures du béton, dans les failles des égouts ou à l'ombre des squats," who "fleurissent les trottoirs quand la nuit est tombée, dénudent leur corps de déesse pour une pièce d'argent, se peignent le visage de couleurs pour cacher leur douleur" and who, "trop tôt déflorées, deviennent une charge dans les tribus" ["the wild flowers blown by the wind, seeking a little light in the cracks of the concrete, in the fissures of the sewers or in the shade of the squatter settlements," who "bloom on the sidewalks when night has fallen, baring their goddess-like bodies for a piece of silver, and paint their faces with colors to hide their pain" and who, "deflowered too soon, become a burden on the tribes"; 23-4].

This political vision has an evident utopian dimension, subtended by Christian principles of peace, love, and forgiveness dear to Tjibaou as a former Catholic priest. Yet *La Parenthèse* does not present a pastoral idyll, but pursues a more complicated analysis that grapples with the uncompromising operations of realpolitik (of which Tjibaou himself was profoundly and painfully aware, and intimately engaged with). Gope's play might be better read as a version of the post-pastoral, which Terry Gifford describes as "a pastoral informed by ecological principles of uneven interconnectedness, as well as an educated understanding of the symbiotic link between environmental and social justice, at both the local level and beyond" (Tiffin and Huggan 115). The play draws attention, through the garden's cultivated and mediated character, to the conscious construction of national and cultural identity—the concept that "belonging is as much created as retrieved" (Tiffin and Huggan 89)—while the Gardener foregrounds the difficult labor involved in maintaining a *sensus communis*. Gope suggests that one interpretation for the Gardener might be the artist, a figure who traverses garden and House, and whose work sows and nurtures ideas and actions that promote harmonious growth (60). Yet Tjibaou's legacy is under threat throughout the play; the Gardener is an embattled character whose plans for the garden are constantly at risk from internal and external forces. Gope thus dramatizes a critical socioecology that publicly exposes the tensions, compromises, battles, and dreams that have characterized New Caledonia since the late 1980s, intercalating the legacies of the past and the hopes for the future with the troubled circumstances of the present.

The Gaiac tree symbolizes what is centrally at stake in this milieu. He[8] represents the Indigenous histories, ancestral wisdom, and customary knowledge that form the epistemological and ontological grounding for a modern Kanak identity that balances past heritage and future planning, and which is most imperiled in expedient reinventions of culture. Gope chose the Gaiac (*acacia spirorbis*, a kind of wattle tree endemic to New Caledonia) instead of another symbolic tree like the coconut, the pine, or the banyan, because of the special meanings and uses of the tree in his local region, and also in response to an actual incident in which several healthy gaiacs at the Cultural Center were uprooted for aesthetic reasons (61). As the Gardener explains, "Gaïac, plante pionnière, je n'entretiens pas seulement l'arbre que tu es, mais ce que tu représentes pour moi, pour nous, toi l'arbre-ancêtre, symbole d'une alliance forte qui unit le sang et la sève, le rouge et le blanc, dans un corps de chair noire, dans un corps de bois vert où l'Histoire s'est écrite" ["Gaiac, pioneer plant, I maintain not just the tree that you are, but what you represent for me, for us; you are the ancestor tree, symbol of a strong alliance that unites blood and sap, red and white, in a body of black flesh, in a body of green wood, where History is written"; 15].[9] The Gaiac is a repository of cultural memory and a source of sustenance for the wounded and the lost, but he is more than an abstract symbol, pointing to the practical interdependence of human and plant realms: "Tu as été poteau de nos cases et tes fleurs ont chassé les mouches qui venaient dévorer notre peau. Ton écorce a tenu ensemble nos paniers, ensembles nos ignames et, sous ton ombrage, le soleil ne nous a pas brûlés" ["You were the posts of our huts, and your flowers drove away the flies that were devouring our skin. Your bark held together our baskets, all of our yams, and under your shade the sun did not burn us"; 15]. This acknowledgement enables Gope to adumbrate an alternative version of natural history that foregrounds Indigenous knowledge in contrast to the imperatives of Western historiography. Nevertheless, the Gaiac is an endangered species in the garden, requiring the Gardner's protection from the hostile emanations of those who "exhalent le parfum

8 Gope has noted that, in general, the gender of the plant characters in the play is not important, but he specifies that Gaiac is male (64). This choice enables Gope to draw attention to particular strengths and vulnerabilities in Gaiac's character (such as his social and leadership responsibilities versus his inability to regenerate without assistance). It is possible that this approach reflects the playwright's own concerns about cultural values couched too deeply in masculinist terms.

9 One Kanak origin myth tells of human beings deriving from trees, specifically the houp tree (*Montrouziera cauliflora*).

de la mort, l'odeur de la corruption, l'odeur qui ruine, l'odeur qui divise, l'odeur qui assassine" ["exhale the scent of death, the odor of corruption, the odor that ruins, the odor that divides, the odor that murders"; 16].

This element is exemplified by Hortensia (hydrangea) who represents settler-colonial interests, especially those invested in right-wing politics and the primacy of Western enterprise. Used to the entitlements granted by French colonialism and threatened by the post-Matignon commitment to Indigenous cultural expression, this character emphasizes how the putative peace plan has engendered new internecine tensions. Hortensia expresses jealousy and anger over the perceived privileges that Gaiac receives: "toi qui vides la terre de l'eau dont nous avons besoin et pour qui on a toutes les attentions" ["you, who empty the land of the water that we need and who gets all the attention"; 16]. The Gardener reminds Hortensia that "je prends soin de toi comme de lui. Mais la terre est pleine de tes semblables, tandis que sa survie dépend de moi" ["I take care of you as I do him. But the earth is full of your fellow species, while his survival depends on me"; 16]. Yet Hortensia's anxiety elaborates real-world concerns about the perceived diverting of status and resources from other demographic groups on the "arbitrary" basis of first presence, and mirrors statements made by representatives of the French Republic against Kanak independence on the grounds of indigeneity. President Mitterand's 1988 comment, for instance, expressed sympathy for Tjibaou's struggle, but made it clear that "I do not believe that historical precedence of the Canaques [*sic*] is sufficient to create rights. History against history: the Caledonians of European origin have also, through their toil, transformed the earth, been nourished by its substance, have put down roots" (qtd in Waddell 152).

But Hortensia goes further to articulate an exclusionary vision for the garden, conveyed through the vocabulary of natural selection. Rejecting the concept of harmonious co-existence as an unsupportable theory (41), Hortensia finds the garden to be "plein d'êtres pitoyables. Il faut le nettoyer" ["full of pathetic beings. It must be cleansed"; 17]. The prime targets for eradication are "ces gros banians […] ces gros gaïacs puants, avec leurs grosses barbes pendantes, avec leurs grosses racines fouineuses […] La vérité, c'est que notre monde est touché par les conséquences de la dégradation que ces plantes pionnières entretiennent sous leurs ombrages" ["these big banyan trees, […] these large stinking gaiacs, with their great dangling beards, with their thick prying roots […] The truth is that our world is affected by the consequences of the degradation that these pioneer plants maintain beneath their shade"; 18]. By framing Hortensia's claims in these terms, Gope cites and ironizes the insidious deployment of Darwinian logic that has frequently informed colonial attempts to naturalize foreign intercession and land confiscation.[10] Hortensia's attempt to usurp mastery of the garden, however, by gathering an army of poisonous and parasitic plants to drive out Gaiac, projects a direct warning about the consequences of social conflict and cultural evacuation if the country remains on its imbalanced trajectory.

Gaiac is not only at risk from dissenting elements among the "grassroots" community, but from the cultural policies dictated by the House. While the Manager—ensconced in his stultifying office full of artificial plants—does not share Hortensia's divisive commitment to ethnic hierarchies, his pragmatic goal of moving Kanak culture forward is based too firmly on an outward-looking and future-oriented vision, rendering Gaiac a superfluous, even embarrassing, hindrance. "Ce jardin," he declares, "a besoin de respirer, d'être peuplé de nouvelles couleurs. Gaïac gêne, il a fait son temps, son règne est fini. La Maison a besoin d'un nouveau souffle, d'être modernisée, réactivée, réactualisée. Le jardin aspire à une nouvelle vie, une nouvelle âme" ["The garden needs to breathe, to be filled with new colors. Gaiac is an inconvenience, he has had his day, his reign is over. The House needs a new lease of life, to be modernized, revived, refreshed. The garden aspires to a new life, a new soul"; 28]. Despite the Gardener's protest that the House will lose its meaning without Gaiac, asserting that "On n'arrache pas sur un coup de tête ce que la nature a mis tellement de temps à élever!" ["One does not tear out on a whim that which nature has taken so long to raise!"; 26], the Manager is obdurate in his belief that the garden should, officially and as quickly as possible, "accueillir des plantes jamais vues" ["host plants never seen before"; 28].

Gope's cautionary parable reaches its apogee when the Manager's contractors, assisted by Hortensia's botanical militia, inevitably uproot Gaiac. This traumatic imposition on the garden is presented as the culmination of a series of wayward decisions made by the inhabitants of a House whose "seule détermination pour notre culture" ["sole determination for our culture"] is now dictated by "l'État français" ["the French State"; 45], and that gives rise not to a new order, but to a

10 See, for instance, Jules Garnier's 1871 argument to justify the establishment of New Caledonia as a free settler colony: "It is evident that the Kanak, like all primitive peoples, live miserably on land that would support with abundance fifty times more European inhabitants [...]; isn't it just that they give way to those who, by patient and active study, have over the course of many centuries, assembled a sum of knowledge and materials that enables them to arrive at this remarkable outcome of being able to draw from the soil fifty times more than the native can, and isn't it also a law constantly sanctioned in the natural world by thousands of events, that the skilled displaces the unskilled, and the strong displaces the weak!" (Garnier 214-15, qtd in Waddell 37).

troubling vacancy. Woman One, who shares the running of the House,[11] reminds the Manager that instead of listening to the garden, "nous avons arraché ce qui nous dérangeait, mais le trou est béant, rien ne pousse à la place" ["we have torn out that which bothered us, but the hole is gaping, nothing grows in its place"; 44]. In these scenes, Gope expresses his criticism of how the potential vested in the Cultural Center specifically, and in Tjibaou's greater vision for the nation, has been corrupted and sold out, with people paying mere lip service to Tjibaou's ideals and memory. Woman Two[12] remarks that, "Ceux qui parlent au nom de la Maison et disent la politique du Pays portent le souvenir du Vieux comme un bandeau sur leur tête. Mais au fond, rien ne subsiste de son histoire" ["Those who speak in the name of the House, and dictate the Country's policies, bear the memory of the Old Man like a bandana on their heads. But deep down, nothing remains of his history"; 49]. After the murder of Gaiac, the House becomes alien, devoid of the cultural roots that connect it meaningfully to the people; as Woman Two avers: "La Maison ne nous ressemble pas, je ne la reconnais pas" ["The House does not resemble us, I do not recognize it"; 48]. Gope presents an uncompromising indictment of New Caledonia's contemporary governance and self-identity, and, as is typical of his work, proffers no easy answers, instead encouraging his audiences to assess their own complicity in the milieu refracted on stage and to discern their own solutions. This generative approach is modeled by Woman One, who admits that, "Si je me suis perdue en chemin, il est de ma responsabilité personnelle de retrouver ma voie" ["If I got lost along the path, it is my personal responsibility to find my way again"; 49].

Similar anxieties and possibilities are expressed by characters beyond the confines of the House. The Gardener registers the chaos and disorientation of the times, lamenting that "Mon jardin est devenu terre de désolation" ["My garden has become a land of desolation"], which the ancestors have abandoned in shame due to a culture that is showcased "comme une prostitueé" ["like a prostitute"], and because of "nos méchancetés, [...] notre égoïsme, ce culte étranger de la personnalité" ["our wickedness, [...] our egoism, the strange cult of personality"; 34]. The Tribal Chorus addresses the spirit of Tjibaou with a vision of a lost, deracinated people more invested in economic advantage than cultural

---

11 Woman One is the widow of an assassinated political leader who takes public office herself after her husband's death. She might be interpreted as an allusion to Tjibaou's widow, Marie-Claude, who became the President of the ADCK after its establishment in 1989.

12 Like Woman One, Woman Two is the widow of a political leader, but chooses to pursue domestic responsibilities after her husband's death. She might be construed as the wife of Yeiwéné Yeiwéné.

identity: "[T]es frères ne rêvent plus de liberté. Cœurs enchaînés, ventres fragiles, ils sont attachés contre le bois mort au cœur de la ville. Elle a lié leurs pieds, leurs mains, et ils parlent un langage que les leurs ne comprennent pas" ["[Y]our brothers no longer dream of freedom. Chained hearts, weak stomachs, they are tied to the dead wood in the heart of the city. It has bound their feet, their hands, and they speak a language that they do not understand"; 53]. Yet, Gaiac's promise that "je repousserai sur un seul grain de ce sol" ["I will grow again on a single grain of that soil"; 34] extends the hope for cultural regeneration and a reinvigorated social ecumene whose impetus might come from below as well as from above.

Gope's challenge to audiences is deliberately not limited to the binary contentions and hierarchal dichotomies of the settler-indigene relationship, exploring other, lateral affiliations interwoven throughout New Caledonia's social fabric. One such group is represented by Geranium, who stands for those people forcibly sent to New Caledonia during its colonial history, "exilés sur cette terre contre notre volonté" ["exiles in this land against our will"; 19].[13] Principally, Geranium evokes the massive wave of indentured laborers (mainly Oceanian), co-opted as plantation workers, nickel miners, ship-hands, domestic servants, and convict escorts, who comprised the backbone of white industry for much of the later nineteenth century and well into the twentieth (Shineberg 187). Geranium bears a violent history of displacement and servitude, compounded by the recognition of his complicity in Indigenous dislocation, but has struggled to find peace and a place on a foreign soil (40-1). Resisting Hortensia's domineering machinations, Geranium repeatedly expresses a dedication to reciprocity and mutual respect that involves commitment from all members of the community. At the hole left by Gaiac, Geranium pledges that "Aujourd'hui, je choisis de panser les blessures de cette terre, les blessures que nous avons causées. Je dis bien nous. Tous" ["Today, I choose to heal the wounds of this land, the wounds that we have caused. I mean us. Everyone"; 40]. It is a credo founded on a vision of "harmonie entre les plantes. Je crois à la cohabitation, au destin commun. Je crois à un avenir serein dans ce pays, et qu'il faut se donner les moyens de construire cette liberté" ["harmony between the plants. I believe in cohabitation, in a common destiny. I believe in a peaceful future for this country, and that we must provide the means to build this freedom"; 41]. Through the character of Geranium, Gope is able to articulate a positive future goal while accentuating the productive dynamism and linked histories between

13 Geranium (cranesbill) is a plant found globally; its particular method of seed dispersion, which consists of beak-like column that springs open when ripe and casts its seeds a considerable distance, might account for its allegorical choice in the play.

diasporic and indigenous populations in New Caledonia, particularly those connections that situate Kanak within a broader network of pan-Pacific allegiances.

Ultimately, it is the Gardener-as-artist who provides the final reinforcement of this social mission. Returning to the garden after his defeat, he gently binds together the fractured Chorus with a rope of gaiac bark in a symbolic gesture of reconciliation and unity (51). During the play's closing scene, which takes place at Tjibaou's tomb, the Gardener vows to sweep away the monument's artificial flowers and break its cold marble, refusing to consign Tjibaou's struggles to the forgotten pages of history books. "Ta tombe," he promises, "ne sera pas la vitrine d'une cause déchue, où des passants viennent murmurer des mots doux avec un cœur dubitatif, sceptique, mais ce tertre où les enfants de ton sang viendront chanter les grandeurs de ton nom" ["Your grave will not be a showcase for a fallen cause, where passers-by whisper sweet words with a doubtful heart, skeptical, but the mound where the children of your blood will sing the greatness of your name"; 53]. *La Parenthèse* is thus a profound and personal statement from Gope as a dramatist. It stakes an investment in his artistic role as a caretaker of cultural memory, who revives, via the theatre's material and ideational processes, the figures and affects of the past to be reconceived through the lenses of the present; who propagates ideas and actions that take root in the microcosm of the theatre to spread to the world outside; and who questions, comments, and provokes in service of a broader and more generous social vision. In contrast to those gardeners who "ne cultive que deux variétés de plantes, l'argent et le pouvoir" ["grow only two kinds of plants, money and power"; 45], the Gardener's final declaration expresses a commitment to the nurturing of the populace: "S'il le faut, j'arracherai toutes les plantes de ce jardin, et à leur place, je planterai des hommes!" ["If necessary, I will pluck out all the plants in this garden, and in their place, I will plant people!"; 54].

In *La Parenthèse*, Gope deploys an ecological grammar to parse pressing social and cultural problems that affect his contemporary national community, accentuating the intimate imbrication of nature, culture, and political power. Neither an elegy on the ravages of Western imperialism, nor a celebratory idealization of the organic symbiosis between Indigenous peoples and the land, Gope's post-pastoral allegory presents an incisive critique of New Caledonia's postcolonial history, tabulating what has been gained and lost since the inception of the Kanak independence movement, and since the Matignon and Nouméa Accords. Produced in a twenty-first-century milieu where the country's "economic interests take primacy over political ones and where cultural concerns are once again condemned to the margins" (Waddell 206), the play asks urgent questions about cultural identity and expression, foregrounding a responsible relationship with the natural world as an integral part of Kanak heritage and as the model for a shared commons where—in the Gardener's words—"chaque élément est faiseur de vie pour l'autre; tout est collectivement fondé sur la complémentarité, sur la complicité" ["each component is the maker of life for another; everything is based collectively on complementarity, on complicity"; 17]. This is an ideal, however, that can only be achieved with hard work. Tjibaou's famous maxim, "Notre identité est devant nous" ["Our identity is in front of us"], points to a national culture that is constantly constructed for the future, and this concept is literalized on the theatrical stage, where its possibilities can be rehearsed and subjected to critical scrutiny.

As Wendy Arons and Theresa May point out, a chief paradox of ecological theatre is that—in the West, at least—"one of the key means of shaping and transforming human values is the arts, but the arts [...] have traditionally been conceived as the activity that most divides humans from 'nature.' Moreover, theater in its present form—with its emphasis on human conflict in the context of human institutions—occupies a space at the far end of that spectrum" (1). Gope's particular form of ecodramaturgy, which mobilizes the natural environment to provide a more lucid view of the operation of human institutions, along with his self-reflexive mediation of the artist as a gardener, suggests ways in which these different domains might be broached. In so doing, Gope's play suggests a wider purview for ecocriticism. Consequently, whereas *La Parenthèse* might at first seem highly personal and locally focused, it bears a wider significance that speaks to mutual concerns about cultural direction and self-determination in other parts of Oceania and the postcolonial world. Bringing the work of this Francophone playwright into translingual dialogue with Anglophone commentary illuminates these connections, and offers a richer conception of Pacific Islands drama.

Works Cited

Apio, Alani. *Kāmau.* Honolulu: Palila Books, 1994.

—. *Kāmau A'e.* Honolulu: Palila Books, 1998.

Arons, Wendy, and Theresa J. May, eds. *Readings in Performance and Ecology.* New York: Palgrave Macmillan, 2012.

Bihan, Anne. *La Nouvelle-Calédonie, ses literatures et ses livres: Bibliographie selective et commentée.* Noumea: 17e Rencontres Francophones, 2012.

Chanwai-Earle, Lynda. *Heat.* Typescript. Wellington: Playmarket, 2010.

De Blas, Alexandra. "Carving out Issues: Interview with Emmanuel Kasarherou, Cultural Director at the Tjibaou Centre." *Culture & Tradition: Tjibaou Centre.* Australian Broadcasting Corporation, 2000. Web. 28 November 2012.

Dorras, Jo. *Mi Wan Totel.* Typescript. Port Vila: Wan Smolbag Theatre, 1995.

Fenoglio, Micaela. *Des racines et des ailes: La Littérature francophone de la Nouvelle-Calédonie*. Turin and Paris: L'Harmattan, 2004.

Garnier, Jules. *Voyage autour du monde: La Nouvelle Calédonie (côte orientale)*. Second édition. Paris: Henri Plon, 1871.

Gobrait, Valérie. *Le Partage de la Terre*. Ed. Sonia Lacabanne. *Théâtre océanien: Anthologie*. Papeete: Au Vent Des Iles, 2011. 87-145.

Gope, Pierre W. *Le Dernier Crépuscule*. Noumea: Grain de Sable, 2001.

—. *The Last Nightfall*. Trans. Baineo-Boengkih and Penelope S. Keable. Noumea and Suva: Grain de Sable/Institute of Pacific Studies, University of the South Pacific, 2001.

—. *La Parenthèse*. Noumea: Traversées, 2005.

—, and Nicolas Kurtovitch. *Les Dieux sont Borgnes*. Noumea: Grain de Sable, 2002.

—. *Gli dei sono Ciechi*. Trans. Micaela Fenoglio. Turin: L'Harmattan Italia, 2003.

Hunt, John Dixon. *Greater Perfections: The Practice of Garden Theory*. London: Thames and Hudson, 2000.

Jouve, Dominique. "The Cultural Impacts of the Matignon-Oudinot Agreements." *Le Réseau Asie*. 2005. Web. 17 February 2010.

Keown, Michelle. "*Littérature-monde* or *Littérature océanienne*? Internationalism versus Regionalism in Francophone Pacific Writing." Ed. Alec G. Hargreaves. *Transnational French Studies: Postcolonialism and* Littérature-monde. Liverpool: Liverpool UP, 2010. 240-57.

Kneubuhl, Victoria Nalani. *Ka Wai Ola (The Living Water)*. Eds. John H. Y. Wat and Meredith M. Desha. *He Leo Hou: A New Voice—Hawaiian Playwrights*. Honolulu: Bamboo Ridge, 2003. 209-66.

Maclellan, Nic. "Conflict and Reconciliation in New Caledonia: Building the Mwâ Kâ." State, Society and Governance in Melanesia Discussion Paper, Research School of Pacific and Asian Studies, Australian National University. Web. 15 January 2011.

Mateata-Allain, Kareva. *Bridging Our Sea of Islands: French Polynesian Literature Within an Oceanic Context*. Saarbrücken: Verlag Dr Muller, 2008.

Nicole, Robert. "Resisting Orientalism: Pacific Literature in French." Ed. Vilsoni Hereniko and Rob Wilson. *Inside Out: Literature, Cultural Politics, and Identity in the New Pacific*. Lanham, MD: Rowman and Littlefield, 1999. 265-90.

Ponifasio, Lemi/MAU. *Birds With Skymirrors*. Performance. Schauspiel Essen Grillo Theater, Berlin. Theater Der Welt, RUHR, 13-16 July 2010.

Shineberg, Dorothy. "'Noumea No Good. Noumea No Pay.' 'New Hebridean' Indentured Labour in New Caledonia, 1865-1925." *Journal of Pacific History* 26.2 (1991): 187-205.

Stevenson, Robert Louis. "The Ideal House." *Essays of Travel*. London: Chatto and Windus, 1905. 199-206.

Takaku, William. *Conference of the Birds*. Unpublished typescript, c.1989.

Taylor, Apirana. *Message in a Bottle*. Performance. Taki Rua Theatre, Wellington, 1994.

Tiffin, Helen, and Graham Huggan. *Postcolonial Ecocriticism: Literature, Animals, Environment*. London and New York: Routledge, 2010.

Waddell, Eric. *Jean-Marie Tjibaou, Kanak Witness to the World: An Intellectual Biography*. Honolulu: U of Hawai'i P, 2008.

Unfolding the Cloth

Patterns of Landscape and Identity in The Conch's *Masi*

Lisa Warrington and David O'Donnell

University of Otago and *Victoria University of Wellington*

New Zealand-based Pacific theatre company The Conch presented the world premiere of their latest work *Masi* at the New Zealand International Arts Festival at Wellington's Te Papa in March 2012. The company, founded by Nina Nawalowalo, a New Zealander of Fijian and English parentage, and her English husband Tom McCrory, is noted for its staged evocations of physical Pacific environments within an intercultural framework. In *Vula* (2002) the company created a visceral environment using natural materials (water, fibres, fabrics, fans) as a conduit for telling stories drawn from the cycles of women's work in an archetypal Fijian lagoon. In their new work, the unifying image is the creation and patterning of *masi*, a traditional Fijian *tapa* cloth, which has a powerful spiritual significance.

The Conch's philosophy is to develop "an entirely new theatrical language capable of expressing and advancing the core values of Pacific Island people to a broad theatre audience." (The Conch https://www.facebook.com/TheConchNZ/info) It is the expression of this "new theatrical language" within an ecocritical context that we explore here. *Masi* takes an ecological stance, embracing the inter-connectedness of art, land, identity, culture, politics, environment and spirituality. Organic nature and stage illusion become active partners in a hybrid dramaturgy that highlights questions of authenticity versus theatricality.

Masi is a complex project, encompassing significant contributions from three countries, and, at the time of writing, its directors still considered the production a work in progress. In preparing this essay we conducted interviews with *Masi*'s co-devisors and directors Nina Nawalowalo and Tom McCrory, their English collaborator the illusionist Paul Kieve, and lead object animator Salesi Le'ota. In addition to attending the premiere performance, we also observed several rehearsals, obtaining further insight into the group devising process. In order to come to a reading of the work that is relevant to the Pacific context first and foremost, we draw on ecocritical and post-colonial theory, including Homi Bhabha's notion of "Third Space" and a significant essay by eminent Samoan writer Albert Wendt.

The Conch: A New Theatrical Language?

The Conch was founded in 2002 by Nawalowalo and McCrory to facilitate the production of *Vula*, in which much of the meaning was conveyed through physical action and dance, with some Samoan and Fijian dialogue.[1] A conch is a seashell used as a musical wind instrument, primarily as a means of calling or summoning. Therefore the company's name immediately conjures the notion of art literally being produced from nature, as well as the summoning of an audience. The verb "conjure" is especially relevant to Nawalowalo's work because she is a trained magician, having worked for six years in the magic industry in Britain, performing and teaching with magician Richard MacDougall. In London she met Paul Kieve, who would later become a key collaborator on *Masi*, and she went on to work the magic circuit in the United States and Europe.[2] The use of magic in *Vula* included the appearance and disappearance of objects through skilled manipulation of woven screens and fans—techniques that would be expanded significantly in *Masi*. Nawalowalo played basketball for New Zealand in the 1980s, and has compared her athletic skills to those required in creating and choreographing physical theatre. She also has a background in mime and mask, having trained with Robert Bennett of Mime International in New Zealand, Desmond Jones in London and Antonia Fava in Italy (Nawalowalo 2011 Interview).

Vula was entirely grounded in a sense of connection to the natural world, through the imagery of water and moon (*vula* in Fijian), and their connection with the ocean. The stage was flooded with several inches of water in which the actors performed ankle-deep, celebrating Pacific women's identity via their daily work cycle in the lagoons which Nawalowalo had observed on her first visit to Fiji with her father.

> In '94 I came back [to New Zealand] because my father wasn't well and I took him back to Fiji [...] I'd come from the States, from doing a season of *The Magic Castle* and it was a massive turning point because we went to Kadavu, and I was with all the women and it's very traditional.[3] I knew

1 Samoan dialogue was introduced because the original cast members (Susanna Lei'ataua, Fiona Collins and Tausili Mose) were Samoan actors. At that time, there were no suitable Fijian actors available in Wellington.

2 Kieve has had considerable success in his field on the West End stage and as the only illusionist working on the *Harry Potter* films.

3 Kadavu is the home island in Fiji of the Nawalowalo family.

> that the time was actually short with my father. And I'd been exploring a very European style of theatre, such as Basle masks, and suddenly I wanted to explore things to do with my father and my heritage. It was a really key turning point. (Nawalowalo 2011 Interview)

Vula thus evolved as an expression of Nawalowalo's desire to explore her Fijian heritage, informed by her knowledge of Western theatre techniques.

While *Vula* was critically acclaimed in Wellington, its reception beyond a local context is important in assessing the wider significance of the work. The Conch's mission to take core Pacific values to a broader audience was fulfilled when *Vula* received invitations to Fiji, the South Pacific Arts Festival in Palau, the Sydney Opera House, Holland and London's Barbican Theatre.[4] The international interest in The Conch is evidence not only of their artistic innovation, but also of a wider recognition of the relevance and timeliness of the company's ecological and environmental themes, as this London review comment suggests:

> Then came dances based on the fauna of the Pacific, with skilfully controlled fans portraying birds swooping and fish swimming in wonderful formations. This also revealed how everyday objects (fans and brooms) are frequently made from the material of nature, although this had a double-edged ring. Did this, as the programme argued, reveal how such objects retain nature's life, or simply illustrate how mankind furthers its ends by destroying nature? (Smith)

This reviewer draws attention to the ecocritical basis of The Conch's work. Not only were the key props in *Vula* Fijian objects constructed from natural materials, the production also led the reviewer to reflect on the impact of human intervention in nature. In her definition of ecocriticism, Cheryll Glotfelty argues:

> In most literary theory "the world" is synonymous with society—the social sphere. Ecocriticism expands the notion of "the world" to include the entire ecosphere. If we agree with Barry Commoner's first law of ecology, "Everything is connected to everything else," we must conclude that literature does not float above the material world in some aesthetic ether, but, rather, plays a part in an immensely complex global system, in which energy, matter, *and ideas* interact. (xix, emphasis in original)

---

4 This international success has been mirrored by MAU, another Pacific physical theatre company based in New Zealand, and led by Samoan choreographer Lemi Ponifasio. Amongst many prestigious invitations, MAU's *Requiem* (2006) was commissioned by Peter Sellars for the Mozart 250th anniversary festival in Venice. Like The Conch, MAU's work often has environmental themes, strongly apparent in their most recent work *Sky Birds with Mirrors*, which was premiered in the same Wellington arts festival as *Masi*.

Glotfelty's extension of Commoner's emphasis on the interconnectedness of the ecosphere to include literature reflects long-held indigenous Pacific views on art and environment. In an influential essay, "Tatauing the Post-Colonial Body," Wendt specifically addresses the relationship between art and the Samoan view of social and ecological reality, the *va*:

> *Va* is the space between, the betweenness, not empty space, not space that separates, but space that relates, that holds separate entities and things together in the Unity-that-is-All, the space that is context, giving meaning to things. The meanings change as the relationships/the contexts change. (We knew a little about semiotics before Saussure came along!) A well-known Samoan expression is "la teu le va." Cherish/nurse/care for the va, the relationships. This is crucial in communal cultures that value group, unity, more than individualism; who perceive the individual person/creature/thing in terms of group, in terms of va, relationships. (88)

This concept of *va*, focussing meaning and interpretation on the spaces *between* elements, has strong similarities to an ecological position which sees nature and humans operating in a complex and ever-changing series of relationships. Using Samoan *tatau* (tattooing) as his example, Wendt argues that the communal foundations of indigenous Pacific cultures create artworks that express the interrelated nature of "the people, the aiga,[5] the village, the community, the environment, the atua[6], the cosmos" (89). These ecological connections inform Pacific artistic production, which is often based on symbolic expressions of the natural world: "Over its long history the basic motifs of the tatau and malu[7] developed out of representations of atua or out of nature—plants, objects, creatures" (Wendt 92). Similarly, Glotfelty outlines the project of ecocriticism in terms of the relationships between humankind and the natural environment: "How do our metaphors of the land influence the way we treat it? How can we characterize nature writing as a genre? In addition to race, class and gender, should *place* become a new critical category?" (xix). Wendt's interpretation of *va* is useful in understanding not only the themes of *Masi* and its predecessor *Vula*, but also for understanding the significance of the journeys and creative relationships that were formed in devising the works.

In *Vula*, the flooded stage radically altered the nature of the theatre space to evoke a sense of place that was explicitly gendered. The production celebrated Pacific female identity through a visceral connection with place, equating essential tasks such as fishing and the

5 Family.

6 God.

7 Female tattoo.

washing of clothes and bodies with the natural cycles of the moon and tides. The women were seen to be working in harmony with nature, the influence of modern industrialised society only briefly hinted at by the presence of a television set. However, this connection with technology was soon interrupted as a thunderstorm forced the women to shelter the TV under a mat. This comic sequence suggested that the forces of nature are more powerful than electronic media, a view that seems more potent when presented from the perspective of indigenous Pacific women who spend all of their day outside, rather than in a built environment. If, as Glotfelty suggests, place should become a new critical category, Nawalowalo's indigenous Pacific theatre creates a dialectic between place and environmental issues that is performative and interactive.

Masi: Staging "Interconnectedness"

Masi celebrates and combines traditional Fijian performance and culture with Western theatre elements, including stage illusion. A memory piece inspired by photographs and objects, it stages the meeting and romance of Nina's parents. This segues into a meditation on the passage of time, and an evocation of Fijian land and culture specifically through the recalling of the spirit of Nina's now-deceased father. This hybrid physical theatre performance (which contains very little spoken text) is filtered through an autobiographical lens, with Nawalowalo herself playing a central role, embodying the cross-over of three cultural spaces—Fiji, New Zealand and England. By placing herself as an observer physically at the centre of the piece, Nawalowalo explores the cultures forming her genetic make-up from the neutral ground of New Zealand, thus sparking the hybrid nature of the work. Her European side is most strongly represented in *Masi* by the use of photography, illusion and stage trickery. She reveals her Fijian self through an evocation of her father's culture, and through the presentation and manipulation—beating, folding, unfolding, flying of *masi* cloth itself.

> On my trips to visit family in Fiji I have been gifted *masi*. They have hung on the walls of my home sitting among the variety of objects and images, key signifiers of my identity, my heritage and heart links to my family. In their paper thinness they have folded into boxes and trunks to travel with me yet when they unfold with them unfolds a greater story, a greater mystery. (Nawalowalo Education Pack *Masi*)

The combination of natural and theatrical elements in *Masi* becomes a performative manifestation of the *va*, foregrounding ecological unity in dialogue with the spaces between elements, leading to fresh interpretations of Pacific identity.

As with *Vula*, the title *Masi* unpacks to reveal a dynamic representation of the land and its people. Like the Samoan *tatau*, the Fijian *masi* cloth symbolizes the land and the natural world. Unlike the *tatau*, however, which is a male tradition, the makers of *masi* cloth are predominantly female. This builds on the feminist themes of *Vula*, foregrounding the work and contribution of women to Pacific cultural identity.[8] *Masi* is made by groups of women from the inner bark of the paper mulberry tree, which is stripped, soaked, beaten and later stenciled or stamped with distinctive patterns specific to different villages and regions.

> Grown on land which is considered sacred, the *masi* cloth is far more than a nicely designed wall hanging that the visitor may see in a handicraft centre. *Masi* contains the spirit of the land it comes from, the tree it was once part of, and it also contains the essence of the women who beat the cloth and decorate it. Very much an integral part of every aspect of traditional life in Fiji, from the cradle mat to the grave, *masi* continues to be part of contemporary Fijian life and culture. (Spicer and Rondo 13)

Thus, the very fabric used in performance brings the spirit and the land of Fiji itself to the stage, both literally and metaphorically, creating a sense of *masi* as an active character in the work's dramaturgy.

The production was staged at Soundings, a proscenium arch theatre space at Te Papa, the Museum of New Zealand in Wellington.[9] The proscenium was framed with *masi* cloth, wrapping or reclaiming the western stage in a distinctive Fijian cultural framework. Like the flooded stage in *Vula*, this re-framing metaphorically claimed the stage as a site where Fijian identity could be performed and explored. In considering the potential of the theatre to represent the natural world, performance eco-critics Sweeting and Crochunis note the implicit paradox: "With its rows of seats in the audience and its performers on the stage, a conventional realist theatrical space seems far removed from the wild" (325). The re-framing of the proscenium with *masi* fabric signals a dialectic between the natural world and the artifice of the theatre. While *Masi* is certainly not a realistic performance, this proscenium arch, end-stage audience/actor relationship was necessary for the magic effects designed by Paul Kieve, which were considerably more complex than

---

8 *Masi* animator Salesi Le'ota was drawn to work with The Conch because it was "very different from all the other [script based, male-driven] Polynesian theatre that was around at the time. The big difference with Nina's work is the female focus of it" (Le'ota).

9 Given in full, the museum's title is Te Papa Tongarewa, literally "container of treasures." However, to unpack the name further, it can mean: "our container of treasured things and people that spring from Mother Earth here in New Zealand" (Te Papa) http://www.tepapa.govt.nz/AboutUs/history/Pages/MeaningofTePapaTongarewa.aspx.

those used in *Vula*. These require wings, precise heights and a certain distance from the audience in order to realize the illusion. As Emma Willis notes, "unlike the process of masi making, whose methods and processes were exposed, Kieve's magical effects relied on concealment and shadow" (41). It is the interaction between theatrical artifice and natural materials that paradoxically brings nature into prominence in the work.

Placed at the front of the space, stage right, were signifiers of the creation of *masi* cloth in the form of a curved convex-shaped log, the anvil (*dutua*) on which the cloth is traditionally beaten with a mallet (*ike*). Behind the anvil sat *masi* maker Ro Miriama Saunayalewa Tubailagi, who was brought to New Zealand from the Fijian island of Vatulele for this performance. Opposite her, stage left, sat Fijian *vakatara* (chant soloist) and percussionist Peni Jeffrey Lala, who initiated the performance by calling the *vucu* (traditional chant) which served to call all the elements to come into being and to summon Nina to return to the *vanua* (land) (McCrory 2012 Interview).

Photo 1. The *vakatara* (Peni Jefffrey Lala), wrapped in *masi* cloth, calls the *vucu*. The image also shows some of the *masi* cloth which draped the stage. Photography Philip Merry.

There was a strong rhythmic potential, partially realised in the performance, to set the percussive rhythms of the *lali* (small Fijian drum) played by Lala and the beating (*samusamu*) of *masi* against one another as a driving heartbeat of *Masi*.[10]

The essential actions of *masi* making were portrayed in playful physical theatre form, combined with elements of stage illusion. Nawalowalo took us into a memory space via a book of family photographs, which were simultaneously projected on a screen behind her. In the first of many magic tricks, she then assembled a small box made of *masi* panels from which she conjured Fijian actor Kasaya Manulevu. Thus, the natural material (*masi*) appropriated for this stage illusion became an active participant in raising memories. Kasaya, like Nina's father Noa Nawalowalo, comes from Kadavu Island, and represented on stage both the memory and the spirit of Fiji (Nawalowalo [b]). Kasaya proceeded to create her own illusion as she conjured a *masi* tree into being. Using a piece of *masi* cloth in a sequence of classic magician's flourishes, she revealed a seedling which appeared to "grow" in several stages into a genuine paper mulberry tree. A piece of bark from this tree was then ceremoniously handed to Ro Miriama to be beaten. Forty small mulberry trees and the *dutua* used in performance were imported from Fiji, along with many pieces of finished *masi* cloth, so that genuine emblems of the land and the culture were physically present on stage. As Spicer and Rondo note, "The Fijian language uses the word *masi* for both the paper mulberry tree and the finished cloth, as if to emphasise the interconnectedness of the raw material and finished products" (30). The communal spirit of *masi* making was represented through Kasaya performing as a *masi* maker, alongside Ro Miriama. As Nina comments "it wouldn't be right to have one woman alone" (Nawalowalo [b]). Although in reality Kasaya is an actor and a cook, not a *masi* maker, this was an opportunity for a young Fijian woman living in New Zealand to connect with her heritage, expanding the central motif of Nina's performative exploration of her cultural identity.

Nina's parents—Noa, a Fijian lawyer, and Mary, an English immigrant—met in New Zealand in the 1950s in the "neutral" space of the Wellington chess club. Two young actors, Victoria Abbott and Alexander Tarrant, performed as the parents, and their meeting—observed from the present by Nina—was represented by a series of filmic "wipes" (sliding screens), manipulated by hidden animators (a technique Nina also employed to considerable effect in *Vula*). The effect of this stage magic is an arresting one, as figures appear to transform or to vanish before

10 According to Spicer: "Every woman on Vatulele Island has her own distinctive rhythm and beat to pound the *lewana* [the inner bark of the tree]. It is possible to know at any time which woman is making *masi* by listening to the sounds…" (35).

our eyes with the passage of the screens. These wipes, in which Mary, Noa and Nina appeared to pass and sometimes miss each other, brought a particular dynamic to this part of the performance, suggesting the passage of time. As a similar type of "conceal/reveal" illusion was employed in the representation of the growth from seedling to tree of the *masi*, the action of the "wipes" also helps to connect the two cultural spaces here presented, Fiji and Wellington. The two actors appeared in a series of still poses—standing, sitting, holding a suitcase or a chess piece; or in motion, walking across the space, then hidden by a passing screen, with the impetus of their action consequently hidden or re-directed. The use of a suitcase as a key prop (which establishes both Mary and Noa as immigrants) connects this section to the earlier Fijian sequence, in which a small suitcase was opened to reveal a miniature version of Nina's home village, complete with a tiny *masi* tree. The third connecting thread was the presence of Nina herself, who moved freely between the worlds of her parents, observing both, handling both a chess piece and a fan made from *masi*.

Photo 2. (L to R) Ro Miriama Saunayalewa Tubailagi and Kasaya Manulevu work with *masi*, observed by Nina Nawalowalo. Photography Philip Merry.

This "Western" sequence of the performance was characterised by a deliberate lack of colour, and by stillness and arrested motion. A series of black and white photographs of Noa, Mary, Nina and her siblings, was projected on the backdrop, or on to the moving screens, and sometimes on to the figures of performers. These "authentic" photographs by celebrated Netherlands-born New Zealand photographer Ans Westra were the inspiration for many of the stage images, and in some instances were directly reproduced by the bodies of the actors. The fixed and monochrome nature of the photographs set the tone for Nina's connection with her European identity.

Photo 3. Noa and Mary Nawalowalo play chess in their Wellington home. Photography Ans Westra.

The key image was of Noa and Mary playing chess together, signified by two chess pieces, one black and one white, held in turn by the actors or by Nina. The striking contrast between "black" Fijian and "white" Englishwoman was playfully linked to the chess pieces, and to Westra's photography. This deliberate evocation of a simple binary opposition was constantly deconstructed through the use of illusion. The growing relationship of Nina's parents was represented by a short, elegant dance sequence, employing a specifically Western choreographic style of lifts, turns and balances. The two bodies of the actors flowed, engaged and entangled.[11] The stop-start, image-based method of storytelling functioned effectively to represent the filters of time and memory, witnessed by Nina as she must have originally experienced it through seeing photographs taken before she was born.

A significant transition brought six Fijian male performers on stage, clothed in traditional *liku vau* skirts, who danced and sang dynamically in

[11] New Zealand dancer/choreographer Daniel Belton is listed as *Masi*'s choreographic advisor in the programme.

a style which provided a complete contrast to the controlled, still images of the earlier sequence.[12]

Photo 4. Two of the members of the Kabu ni Vanua Dance Group performing in *Masi*. Photography Philip Merry.

Where Nina's parents met in contained spots of light and shadow, these dancers filled the space with movement and high energy, while the stage lighting expanded to brightly colour the space.[13] The performance of a "traditional" dance which might normally be seen in the context of a tourist show was dramaturgically re-contextualized by its integration in this hybridised physical theatre piece. Like the physical presence of the "real" Fijian *masi* maker and *vakatara*, the physicality of the Fijian dancers built the sense of authenticity and the production's staging of nature and environment. Lemi Ponifasio has spoken of the connection between the body and the environment, which extends Wendt's definition of *va*: "We must conceive a body as rooted in the body of felt experience or the body of 'nature.' Here in the Pacific, the notion of the *va* is the lived and cognizant body relationship with the world. The *va*, or the body's intentionality is the foundation of all expression, including dance" (Ponifasio qtd. in Taouma: 140). Ponifasio thus reflects through dance the

12 Nawalowalo auditioned many dancers in Fiji, including some from quite remote localities, but in the end selected six members of the Kabu ni Vanua Dance Group from Suva. Their dance choreographer was Master Laisiasa Veikoso of Suva.

13 The lighting designer was Nik Janiurek. Stage design was by Tracey Monastra.

interconnectedness of body, nature and artistic representation. McCrory sees the dancers as essential collaborators in devising the work, vital to the "authenticity" of the process: "They represent a force which comes out of the *masi* to reach out to Nina, and to help her establish the authenticity of who she is too. They represent the chiefs of Nina's father's line who come to take his spirit when he dies" (McCrory 2012 Interview). For Nawalowalo, the dancers' bodies in the stage space specifically re-invoke her father. "There's so much about their physical being which is just is like my dad, from their calves to their bone structure" (Nawalowalo 2012 Interview).

Noa, dressed in a Western overcoat and suit, was gradually conjured back to his childhood roots in a movement piece involving the waving of large pieces of white cloth, representing both "windy" Wellington and the *masi* of his Fijian heritage.[14] This sequence demonstrates the use of objects and bodies to actively characterize nature. Nina sees it as a "releasing of something in oneself" (Nawalowalo [b]). Manipulated by the dancers, the cloths virtually performed a dance of their own, interacting with Noa. The initial inspiration for this sequence was a photograph of Noa on Wellington's Oriental Parade, where his coat is whipped by the wind—an image that was projected onto the moving white cloths (Nawalowalo 2012 Interview). Time moved backward to reveal Noa as a child in Fiji (played by Jerry Vakamoce), wearing a *lavalava*,[15] who again segued into Noa the adult, now interacting with the Fijian dancers.

At one point in the narrative, several large fans made from *masi* cloth took centre stage in a sequence where the fans, manipulated by invisible animators, appeared to dance and fly of their own accord. This, for Nawalowalo, is the key sequence in *Masi*, bringing together as it does all of the elements that the show represents: "the traditional, the theatrical, the magical and the spiritual" (Nawalowalo [b]). Later, Noa appeared wrapped in white *masi*, signifying his passage towards death, as white *masi* cloth is traditionally used to wrap the bodies of chiefs.[16] The directors staged a stylized representation of Noa's funeral, where a coffin was covered with a large piece of *masi* and by means of one of Kieve's grand illusions made to disappear. This linked together many of the preceding threads of narrative—appearance and disappearance of bodies from boxes, the "cradle to grave" aspect of *masi*, and the element of illusion binding together much of the piece.

---

14 The city of Wellington is known for its frequent strong gusts of wind.

15 A lavalava is a piece of fabric worn as a skirt by both women and men in many Pacific Island cultures.

16 White masi is also used at birth to wrap the new born, and at weddings.

Masi: "The Land is the Soul of a People"

For Nawalowalo, *masi* represents a culture and a people, "because it comes from the earth—which is the land—which is the most precious thing. [...] The land is the soul of a people" (Nawalowalo 2012 Interview). Nawalowalo's artistic exploration of this direct co-relation of land and culture has been fueled by several visits to Fiji, beginning with the trip with her father in 1994 that led ultimately to the creation of The Conch. Taking *Vula* to Fiji in 2006 was a deliberate strategy to build up trust with Fijian authorities, key chiefs (including members of her own family) and government officials. "They have to see it with their own eyes, in order to trust that they will give you traditional knowledge" (Nawalowalo 2012 Interview). When it came to creating *Masi*, Nawalowalo and McCrory needed to seek permission from (amongst others) village chiefs of Vatulele, and the Minister of Cultural Heritage:

> There's very powerful legal protection around traditional knowledge in Fiji. [...] because the indigenous population has made the law to protect the people and the culture. [...] You have to go through the proper protocols to gain permission to use their patterns, for example. To present their patterns on stage we had to have permission from the chief of the village where Ro Miriama comes from. (McCrory 2012 Interview)

Further measures to ensure that cultural protection was fully in place included taking Kieve and composer Gareth Farr (who recorded Fijian choral music which was incorporated in the production) on subsequent trips to Fiji. As an Englishman having his first experience of Fijian culture, Kieve was struck by the reciprocal nature of Western stage magic and *masi* making, which stand as equal partners in the finished production. He gave some small magic performances in Fiji as a form of cultural exchange, including a piece entitled "The Torn and Restored Newspaper":

> I thought if it meant something to anyone it was going to mean something to *masi* makers. They're very familiar with what it takes to fuse strips of material together. It was one of the themes that Nina was interested in—a restoration theme, which is one of the major themes in magic, the idea that you take something that is in fragments, and you restore it back to its original form. And it's extraordinary to think that *masi* comes from the tree, so in a way it's a form of transformation. (Kieve)

Kieve's observations highlight the process by which the intercultural and environmental themes of *Masi* grew from making connections between Fijian cultural production and theatrical transformation.

Nawalowalo's journeys to Fiji were not only a strategy to develop intercultural collaborations between indigenous practices and Western theatre artists. For her as a New Zealander of Fijian ancestry they also represent a tangible homecoming. Like the use of natural materials onstage in *Masi*, Nawalowalo's exploration of her identity connects her sense of home with an ecology of place. Lavery and Whitehead define the connection between ecology and home: "The etymology of the word 'ecology' fuses the meanings of two Greek words: *oikos* (house) and *logos* (science, law). Ecology can be regarded as a type of domestic science, a way of being at home" (111). Through the symbolism and materiality of the *masi* cloth, Nawalowalo has found a metaphor that literally evokes the *oikos*. Fijian houses are filled with *masi*, for both decorative and practical purposes. At the same time, the *masi* symbolizes the earth, both through its transformation from tree bark into fabric, and through the symbolism of the patterns stenciled on the cloth. As a form of clothing, *masi* connects strongly to the body, which Lavery and Whitehead argue is also essential to an understanding of ecology:

> Home doesn't start with language [...] it starts with the body. The body is an amazing ecological resource [...] [it] is the 'first home' and the place or the territory where we live is a 'second home' [...] The 'third' home is the home you discover when you start interacting with the assemblage of body and environment to produce something new. (114)

In *Masi*, indigenous bodies—specifically the six Fijian male dancers and *masi* maker Ro Miriama—interact with environmental signifiers (primarily the *masi* cloth itself) in order to create a new sense of "home" that is explicitly rooted in Pacific connection to the earth, in the same way as *Vula* explored the connection with the ocean. Lavery and Whitehead's notion of a "third home" where body and environment interact recalls Homi Bhabha's concept of a "Third Space," where cultures meet:

> The production of meaning requires that these two places [the "I" and the "you"] be mobilized in the passage through a Third Space. [...] It is that Third Space though unrepresentable in itself, which constitutes the discursive conditions of enunciation that ensure that the meaning and symbols of culture have no primordial unity or fixity; that even the same signs can be appropriated, translated, re-historicised and read anew. (36-37)

Nawalowalo uses the performance to make sense of her identity, seeking to stage the "unrepresentable," those encounters defined by her own relationship with the three "spaces" of being Fijian, British and New Zealander. As her parents were both new arrivals to New Zealand, in a sense they were equals on a foreign soil. Willis notes:

> What is so interesting about the project [...] is how to make a Fijian space outside of Fiji, how to engineer a hybridized space that is porous, fluid

> and energised by the interplay of continuous transformation and continued tradition. In short, how to bring about a dramaturgical alchemy that transports its audience to a new place? (42)

By exploring her cross-cultural genealogy through performance in *Masi*, Nawalowalo enters Bhabha's contested Third Space of representation, re-performing the meeting of her parents in Wellington as a Third Space cultural encounter.

The performance of *Masi* does not strictly follow the oral traditions of the Pacific, instead deliberately reducing spoken text—reflecting Lavery and Whitehead's point that "home doesn't start with language." (114) Apart from a brief introduction by Ro Miriama and the singing and calling in Fijian, the voice plays a secondary role to physical story-telling. By minimising the spoken element, The Conch establishes their desired "new theatrical language," drawing upon both Pacific story-telling traditions and European dramaturgy to create a "Third Space" of meaning.

The hybridity of Nawalowalo's own identity results in a hybrid performance composed of Fijian elements including dance, music and *masi*-making mixed with European elements including magic, physical theatre and photography.

Photo 5. A silhouette of Nina Nawalowalo appears in front of a projected image of her parents. Photography Philip Merry.

For Bhabha, hybridity is to be celebrated, a liminal space that "initiates new signs of identity, and innovative sites of collaboration and contestation" (1). The innovation in Nawalowalo's work resides in her ability to foster cross-cultural collaborations creating striking imagery that can contribute to an understanding of pan-Pacific identity. In *Masi*, Nawalowalo foregrounds the physical presence of the performing Fijian body, dancing, singing and making the cloth. Like Ponifasio, Wendt also celebrates the blending of different cultural influences in terms of the body: "What is the Post-Colonial Body? It is a body 'becoming,' defining itself, clearing a space for itself among and alongside other bodies" (98). Nawalowalo sites her own body at the centre of the performance, playing herself, "becoming" herself through her live, corporeal interaction with the Fijian performers, defining herself and the "Third Space" that she occupies.

Masi: Authenticity and the Staging of Nature

Tom McCrory feels confident that *Masi* was "the most authentic artistic journey that I've ever been on, because I'm working alongside an artist who is actually on a genuine process of journeying into herself." (McCrory 2012 Interview) However, authenticity is a problematic concept. Christopher Balme has examined the tension produced by "the apparent antithetical nature of the terms 'authenticity' and 'theatrical'" (47). In their desire to stage nature, The Conch clearly took great pains to incorporate "authentic" performers and artefacts in *Masi*. But in creating the aesthetics of the performance, several compromises became necessary. Although the *masi* materials and the wooden log/anvil were genuine imports from Fiji, the *masi*-making process seen on stage was of necessity a short-hand theatrical representation of the real thing. The original intention had been to show the processes by which the bark is stripped, beaten, pulped and made into cloth, but it was quickly discovered in rehearsal that this time-consuming practice was impractical to show in a literal sense during performance.

In order to facilitate the stage illusions, some mixing of *masi* cloth from different regions was also required. Each village and island where *masi* is made creates its own distinctive patterns, and it became necessary to use *masi* from locations as diverse as Vatulele, Moce and Nayau, in order to achieve particular textures that could be easily manipulated on stage. For example, *masi* cloth of a very particular pattern which was specifically imported to dress Noa's coffin, could not be used in performance, as it was too new and therefore too stiff to facilitate the intended illusion. In its place, it was necessary to employ a less "authentic" *masi* cloth that had been softened through extensive use in the rehearsal room (Nawalowalo

2012 Interview). This substitution would have been apparent to Fijians in the audience, as is confirmed by animator Salesi Le'ota: "The patterns are so distinct. It's like a tattoo, you can tell where it's from" (Le'ota). Similarly, for the wind sequence, rather than *masi* cloth, another synthetic material called *tyvek* was used, because "it was so beautiful and fluid and light [...] a better fabric to work with" (Nawalowalo 2012 Interview)—and thus provided a more compelling theatrical effect.

Clearly, such artistic compromises needed to be made for practical reasons. This process extended to the desire to use "authentic" Fijian dancers in the piece. Nawalowalo auditioned widely in Fiji (including Kadavu, her father's home island) to find six dancers, but ultimately she felt that most indigenous performers would struggle with the Western theatrical devising process. Therefore, the dancers were selected from the Suva-based Kabu ni Vanua dance group, who were highly experienced performers for Fijian tourism, used to travelling and giving cultural performances around the world. In reviewing *Masi*, John Smythe paid particular attention to the theatrical impact of these dancers: "Of course no evocation of Fijian culture can be complete without a strong dose of masculinity which comes with the Kabu ni Vanua dance group [...] whose contributions are lyrical, muscular, mesmeric and pounding." The critic here makes an assumption that the Fijian dance is an accurate conduit for cultural authenticity. Willis in turn notes, "when the warriors performed their most significant dance sequence my gaze began to feel touristic" (41).

We argue, however, that Nawalowalo's recontextualisation of cultural dance within a hybrid dramaturgy avoids the traps of tourist performance. There is a delicious irony in the use of illusion and magic tricks in a production that appears to foreground cultural authenticity, drawing attention to the slipperiness and contradictions in defining cross-cultural identity. The "authenticity" of the cultural performance contrasts with the trickery of the magic, through which performers and objects appear and disappear, and objects float through the space apparently unaided. The dramaturgy interweaves the solemnity and beauty of indigenous ritual performance with the playfulness of magic and illusion. Such playfulness is common in indigenous Pacific theatre forms with their roots in satiric clowning traditions.[17] In analyzing two performances at the Polynesian Cultural Center in Hawai'i, Balme notes the "playful deconstruction of the tourist gaze with its expectation of authenticity in respect to Polynesian culture" (176). Similarly, *Masi* creates a playful, trickster-like engagement with its audiences, deliberately engaging with this tension

[17] See, for example: William E. Mitchell, ed. *Clowning as Critical Practice: Performance Humor in the South Pacific.*

between authenticity and theatricality. The staged encounters between signifiers of "authentic" Fijian culture and theatrical magic represents a performative interpretation of the Third Space of cross-cultural Pacific identity, symbolised by Nawalowalo's personal search for her cultural homeland. Speaking of her connection with *masi* cloth, she reinforces the connections between cultural identity and the natural world:

> In their fibres, they contain the essence of the land itself, in their dyes and images I could sense the stained fingers and hear the voices of the women working. One day looking at the patterns in the sunlight I recognised in the *masi* a map back to my own cultural roots, which as a Fijian New Zealander I feel, but are likewise mysterious to me. Like the masi tree itself, they sunk into the dark warm soil of Fiji. I felt that if I traced the paths indicated by these fibres I might find my way back. To a time and place. (Nawalowalo Education Pack *Masi*)

Conclusion

In *Masi*, The Conch has created a sophisticated dramaturgy that stages nature in relation to cultural identity in the Pacific. The performance is informed by Nawalowalo's journeys to Fiji to source collaborators and cultural artefacts which are transformed theatrically to create new meanings onstage. Acutely aware of the contradictions between "authenticity" and "theatricality," Nawalowalo employs a playful strategy to stage an ecological encounter with the Pacific. The agency of natural materials transformed into culturally significant artefacts interacts with Western magic traditions. The productive tension between Polynesian illusion and Western magic enacts the "unrepresentable" Third Space of cultural and environmental encounter. Through the symbolism of the *masi* tree and *masi* cloth, nature and culture are seen to interact in a Third Space encounter that reflects the Pacific concept of the *va*, the sense of intricate connection between environment and human cultures. Through experimentation with ecocritical performance, The Conch are realising Glotfelty's notion of "place as a new critical category" and working towards a new theatrical aesthetic which is relevant far beyond the Pacific region. From the beating of bark fibres to the unfolding of finished cloth, *Masi* actively characterises land and landscape within an innovative intercultural dramaturgy.

Works Cited

Bhabha, Homi K. *The Location of Culture*. London and New York: Routledge, 1994.

Balme, Christopher. *Pacific Performances: Theatricality and Cross-Cultural Encounter in the South Seas*. Houndmills: Palgrave Macmillan, 2007.

Glotfelty, Cheryll. "Introduction: Literary Studies in an Age of Environmental Crisis." Ed. Cheryll Glotfelty and Harold Fromm. *The Ecocriticism Reader: Landmarks in Literary Ecology*. Athens, GA: U of Georgia P, 1996. xv-xxxvii.

Glotfelty, Cheryll and Harold Fromm, eds. *The Ecocriticism Reader: Landmarks in Literary Ecology*. Athens, GA: U of Georgia P, 1996.

Kieve, Paul. Interview with David O'Donnell and Lisa Warrington. Soundings Theatre, Te Papa, Wellington. 14 December 2011.

Lavery, Carl and Simon Whitehead. "Bringing it All Back Home: Towards an Ecology of Place." *Performance Research: A Journal of the Performing Arts* 17.4 (2012): 111-19.

Le'ota, Salesi. Interview with David O'Donnell and Lisa Warrington. Aro Café Wellington. 3 March 2012.

McCrory, Tom. E-mail to Lisa Warrington. 22 June 2012.

—. Interview with David O'Donnell and Lisa Warrington conducted via Skype. 27 January 2012.

Mitchell, William E., ed. *Clowning as Critical Practice: Performance Humor in the South Pacific*. Pittsburgh: U of Pittsburgh P, 1992.

Museum of New Zealand Te Papa Tongarewa. Web. 5 June 2013 http://www.tepapa.govt.nz/AboutUs/history/Pages/MeaningofTePapaTongarewa.aspx.

Nawalowalo, Nina. Interview with David O'Donnell and Lisa Warrington. Soundings Theatre, Te Papa, Wellington. 14 December 2011.

—. Interview with David O'Donnell and Lisa Warrington. Toi Whakaari, Wellington. 15 June 2012.

—. New Zealand International Arts Festival Education Pack *Masi*. Web. 29 November 2012. http://2012.festival.co.nz/yk-files/7839c5be584220a9fcd3f463c312e9f7/Masi.pdf.

Smith, Sam. Review of *Vula* at the Barbican Pit London, April 2008. Web. 20 November 2012. http://www.musicomh.com/theatre/vula_0508.htm.

Smythe, John. "The Magical Experience of Discovery and Recovery." Review of *Masi. Theatreview* 3 March 2012. Web. 28 November 2012. http://www.theatreview.org.nz/reviews/review.php?id=4574.

Spicer, Catherine and B.B. Me Rondo. *Fiji Masi: An Ancient Art in a New Millennium*. Queensland: Burleigh Heads, 2004.

Sweeting, Adam and Thomas C. Crochunis. "Performing the Wild: Rethinking Wilderness and Theater Spaces." Ed. Karla Armbruster and Kathleen R. Wallace. *Beyond Nature Writing: Expanding the Boundaries of Ecocriticism*. Charlottesville: UP of Virginia, 2001. 325-40.

Taouma, Lisa. "Getting' Jiggy With It. The evolving of Pasifika dance in New Zealand". Ed. Sean Mallon and Pandora Fulimalo Pereira. *Pacific Art Niu Sila: The Pacific Dimension of Contemporary New Zealand Arts*. Wellington: Te Papa Press. 2002 133-45.

Wendt, Albert. "Tatauing the Post-Colonial Body." *SPAN: Journal of the South Pacific Association for Commonwealth Language and Literary Studies* 62 (December 2009): 83-101.

Willis, Emma. "Enter Here: Emergent Dramaturgies & the Civic Role of an Arts Festival." *Illusions* 43/44 (September 2012): 38-44.

Cleansing the Tapu

Nature, Landscape and Transformation in Three Works by Māori Playwrights

Hilary HALBA

University of Otago

Introduction

How can theatre, that most anthropocentric of artforms, "wedded," as Christopher Manes puts it, "to the monologue of the human subject," (24) be "read under the sign of nature"?[1] Three contemporary New Zealand texts—in the widest sense of the term—by Māori authors: the plays *Woman Far Walking* by Witi Ihimaera and *When Sun and Moon Collide* by Briar Grace-Smith, and the film *The Strength of Water*, written by Briar Grace-Smith and directed by Armagan Ballantyne, will be used to investigate this question.[2] Cognisant of Jodi Byrd's concern that postcolonial theories to do with Indigenous knowledges "deconstruct [these knowledges] into another colonialist discourse," I seek to view these concepts through the lens of an Indigenous epistemological system, matauranga Māori (Māori knowledge) (91).[3] I examine ways in which these texts depict the world of nature and its kinship with human history, social systems and the self. Given the scope of this essay, a complete analysis of matauranga Māori would be impossible, but certain core elements will be exemplified.

In Te Ao Māori (the Māori world), human history and its relationship with the world of nature are inscribed into the self by way of whakapapa (genealogy); whakapapa charts histories of lineage—of the cosmos, elemental deities, humans and the biosphere, with the world seen as a

1 This quote is drawn from the evocative title of the collection of essays edited by John Tallmadge and Henry Harrington.

2 Witi Ihimaera is "of Te Aitanga-a-Mahaki descent" while Briar Grace-Smith belongs "to the Ngati Hau hapu of Nga Puhi" (http://www.bookcouncil.org.nz).

3 Translations from Māori to English are sourced from the Ngata Dictionary on-line, unless otherwise stated.

"unified whole" (Klein 105). Ka'ai and Higgins assert that "Māori interpret the landscape differently from Pākehā and bestow importance on places and geographical features in a different way," and that landscape and languages express "the whakapapa link to the people's origins" (13). In each of the three texts analysed, the history of the human subject interweaves with the world of nature, inextricably linking people with other forces in the ecosystem.

In the personified Māori universe, the land and other elements of the natural world hold their own mauri (life-force) and tapu (restriction or protective prohibition, influence of ancestral entities [cf. Ka'ai and Higgins 18]), and the texts embody this idea. Ka'ai and Higgins define "mauri" as a "life force or life principle and ethos of all objects both animate and inanimate within the universe" (18). Everything in the universe holds its own mauri. In his stimulating essay "Nature and Silence," Manes encourages his reader to consider the viewpoint of cultures "that see the natural world as inspirited," in which "animals, plants, and even "inert" entities such as stones and rivers are perceived as being articulate and at times intelligible subjects, able to communicate and interact with humans for good or ill" (15). Although he uses the contested and problematic term "animistic" to describe cultures that primarily construe the world in this way, Manes nonetheless could be advocating for matauranga Māori. Michael P. Shirres asserts that in the Māori universe "the two worlds [the material and the spiritual worlds] are closely linked with each other" (34). In traditional Māori creation narratives, primal parents Papa-tūā-nuku (earth, female principle) and Ranginui (sky, male principle) spawned—amongst other things—the forests and its creatures, cultivated lands, and the deities responsible for these areas, as well as human beings. The complex, biodiverse matrix of human, spiritual and natural elements in Te Ao Māori underpins this essay's theoretical framework. Landscape ecology "avoids distinctions between natural and disturbed regions" and "accepts chaos theory and its emphasis on diverse complexity" (Howarth 76). Similarly, I argue that natural forces penetrate the "fence of civilisation," that any nature/human delineation remains blurry at best, and the mauri of elements in the non-human world remain palpable (Duerr 30).

A Māori conceptualisation of past and present is coined as "i ngā wā o mua" and has people moving into the future facing the past. However, Māori narratives of creation reproduced in moteatea (folksongs, laments), waiata (songs, chants) and other performance modes render palpable the living universe that was manifest before human emergence. These performative frameworks chronicle the nascence of each constituent part of creation's "dynamic movement" and bring them into the polyphonic present, wherein the narrator connects himself/herself to these non-human elemental and spiritual forces (Shirres 16). If, then, the non-human world is capable of such "utterance," I examine ways in which the playwrights analysed in this essay depict such utterances, and represent the mauri and tapu of natural elements. In addition to focusing upon human kinship with the natural world, a Māori worldview undergirds all three texts, and hence I argue that, in each of the texts, humans are treated as part of the ecosystem rather than as "rational sovereign of the natural order" (Manes 21).

However, to say merely that humankind forms part of nature posits an overly simplistic reading of human impact upon ecosystems, and can provide an excuse for human desecration of the environment. Lynn White Jr. claims, "we shall continue to have a worsening ecological crisis until we reject the Christian axiom that nature has no reason for existence save to serve man" (14). Glen A. Love supports White's claim by cautioning against "[a] notion that human beings are so special that the earth exists for our comfort and disposal alone" (229). In a tangential vein, Frederick Turner argues that humans' "special" standing means that we must "*take responsibility for nature*" and not abdicate responsibility for "the world we have been given to look after" (Turner 50, emphasis in original). Shirres provides a Māori lens through which to consider this point. He quotes the whakataukī (proverbial saying):

> Huutia te rito o te harakeke, kei hea te koomako e ko?
> Kii mai ki ahau: He aha te mea nui o tea o?
> Maaku e kii at: He tangata, he tangata. (53)
> [If you pluck out the heart of the flax bush, how can the bell bird sing?
> You ask me: What is the greatest reality of the universe?
> I reply: The human person!] (53; Shirres' translation)

On the face of it, the whakataukī seems anthropocentric. However, Shirres posits a counter-reading in which the whakataukī "expresses our fragility, our dependence on other parts of creation"; the bell bird cannot sing without the flax plant's sustenance (53). Perhaps we humans might re-imagine ourselves as one (arguably disposable) cog in the world's ecosystem? Can we really think like mountains, to use a phrase John Seed, Joanna Macy, Pat Fleming and Arne Naess use as the title of their book (Seed *et al.*)? Of course, our experience of nature is filtered through human understanding. However, each of the three texts under investigation provides what could be called a rejoinder from nature, although they do not give ready answers or explanations to questions about the quintessential aspects of nature they consider. Ecosystems are capable of great feats of transformation: the planet's geological and atmospheric changes over eons, metamorphoses of form in species from one epoch to the next, shifts in the substance of elements in our universe which we, arguably sapient apes, can neither perceive nor fathom. Yet everything in the Māori universe holds its own tapu and, as Shirres points out, "there is a constant [dynamic] meeting of *tapu* with *tapu*" (37). *Woman Far Walking*, *The Strength of Water* and *When Sun and Moon Collide* explore this viewpoint in different dramaturgical or cinematic ways, and all illustrate their roots in matauranga Māori.

The Past Continuous: Time, People and Land in *Woman Far Walking*

Witi Ihimaera's *Woman Far Walking* depicts a palpable change in Aotearoa's life-essence and a shift in its ecosystem, stemming from the coming of the Pākehā (non-Māori New Zealanders). Before these "goblins" (Ihimaera 10) changed the world forever, "the sky was a man and the earth was a woman" (Ihimaera 9); humans acknowledged their relationship with, respect for and interdependence with these primal parents. *Woman Far Walking* depicts a syncretic universe in which a harmonious congruence between humans and other eco-systems is evident but in which humans—especially colonisers—are simultaneously seen to "tame" the natural world around them. Lynn White Jr. argues: "What people do about their ecology depends on what they think about themselves in relation to things around them. Human ecology is deeply conditioned by beliefs about our nature and destiny" (9). Ihimaera's play portrays "ceremonial time," to use John Hanson Mitchell's term, wherein past and present collapse into layers of memory and challenge chronometric temporality. The memories Tiri both wishes to recall and longs to forget etch themselves into the palimpsest of the landscape. At 160 years of age, Tiri is "an aberration. A freak" (Ihimaera 9); she remains "high and dry like a waka [canoe] in a museum where there are no seas" (13). She cannot die, but endures as a living symbol, named after Te Tiriti o Waitangi (The Treaty of Waitangi[4]) and born the day it was signed (Ihimaera 1). The play's other major character, Tilly, acts as an interlocutor. In his introduction to the playtext, Ihimaera describes Tilly as "a guest at Tiri's wedding"—like the wedding guest in Coleridge's "Rhyme of the Ancient Mariner"—who prompts and challenges Tiri's rememoration. Because the memory of an event is unstable, Tilly provides the nexus of that instability, a fissure in Tiri's memory (3).

A shift can be charted in Ihimaera's play from the sacred world of the traditional weavers, living in a tūturu (wholly) Māori world in which the elemental forces are called daily by their karakia (prayers, incantations), to a torn land after Pākehā incursion. Tiri says of the old weavers,

[4] A founding document of Aotearoa/New Zealand, signed on 6th February 1840 between Māori chiefs and the British Crown.

"[t]hey called the rainbows over the valleys, through the mist and into the sacred work so the aho tapu, the sacred thread, would be in the weft of the weave!" (22). Later, she charts the ecologically and socially destructive effects of the Pākehā invasion: "More pale people, coming down the road. Settlers. Pale faces. Putting up fences. Pale riders upon pale horses. Towns sprouting up all around us. Pale, pale, faces. Replacing our godsticks with their surveyor's pegs" (32). Yet the valley of the rainbows remains a touchstone and Tiri's relationship with that land persists as an elemental and continuative force. She says of her return to the valley:

> I stooped to the ground and I picked up that sweet soil in my hands and held it to the sky and made my karanga to all the rainbows in the world and the people of the valley came to me [...] Haramai koutou ngā kahukura o te ao, haramai, haramai—
>
> [...] And when the rainbows were all in in attendance, all gathered there, I made a pact to them—
>
> For as long as I breathe, and as long as I live, this valley and this land is ours and I will fight to keep it forever and ever— (53)

Ihimaera's stage direction reinforces the valley's healing power: "*Birds begin to sing in a dawn chorus which rises in a crescendo. The sea washes gently on a beach. There is a sense of life renewing itself*" (53). This is not to say that the play revels in a romantic, "primitive" pre-colonial "Eden." The worst excesses of human greed and cruelty to both land and humans are depicted by Ihimaera in his representations of inter-tribal wars—the "musket" wars—and of Pākehā incursion, war and disease. The relationship between war and landscape is imbued with an indelible tapu, a sacredness that cannot be removed, through a litany of battles and pilgrimages. Christopher Manes, quoting Hans Peter Duerr, claims, "people do not exploit a nature that speaks to them" (16), and draws on what he calls the discourse of "ontological humility" (15) where the natural world is treated as a silenced subject (16), often in conflict with economic "progress." *Woman Far Walking* charts more destruction at the hands of Pākehā—to whom nature is evidently "silenced"—as time moves forward. Tilly says, "[...] we pilgrimaged to the Ureweras. [Biggs] attacked us with 140 soldiers at Paparatu. He attacked us again at Te Koneke. As we traversed the Ruakituri, Biggs tried to stop us there also [...]" (37-38). Tiri and Tilly enumerate the deaths in battle of Tiri's children, Hiria and Kereopa, which represent the dark details "by which she marks her history" (49). Tiri's son, "the pale child Pirimia" (61), compounds colonial transgressions as he was conceived after Tiri's rape (93-95). Tiri has a premonition of disaster about him, yet he does not die alongside his brother at the Somme during World War I. Instead, he comes home, bringing with him deadly influenza that decimates almost everyone except Tiri. She laments, "[a]ll my loved ones gone. And I am still here, alone, the last of my kind, and desolate on the shore" (71). Tiri survives and continues to survive, counteracting nature. When the French government tests nuclear devices on Mururoa Atoll in the Pacific Ocean, she recalls:

> When my grandmother saw Captain Cook sailing into Tolaga she said he looked like a goblin. Wherever he goes he murders people [...] He even murders his own. What kind of goblin does that? Now he murders the land and the sea. (76)

Even Tiri's own descendant, Fred, sells the rights to ironsand in the valley of the rainbows to Japanese mining interests (90). Humans pillage Papa-tūā-nuku again and Tiri fears that she will "end up in Japan" (92) if she is interred in the valley's sands when death finally comes, or "as a fish fillet" (2) if she is buried at sea.

The notion of continuance accompanies Tiri's death at the end of the play. Her mokopuna (grand daughter, descendant) Jessica takes over from Tiri's performance of "Ka Panapana," the women's haka, with Jessica's voice "coming out of the future" (Ihimaera 100). Tiri has taught Jessica traditional values that are symbolically encapsulated into the women's haka. These values will continue and Tiri's transformation into a spirit is complete. She "*listens to the voices, her face expressionless and eternal*" like a memory (100). The landscape of Tiri's memories provides a potent site upon which to consider history and the interpenetration of humans and place. Viewing the world through a Māori lens, Tiri both transforms and is transformed by the world she inhabits and with which her sense of self is inextricably linked.

The Strength of Water

Marc Maufort argues that Briar Grace-Smith uses tropes of Magical Realism to tell stories of the Māori world, characterising her dramaturgy as "a dramatic equivalent of postcolonial magic realism" (248). In Grace-Smith's work, aspects of the human, natural and elemental worlds are "re-sensed" through the chimerical grafting of these different ways of knowing the world. The daily and the extra-daily sit side-by-side, intersecting and interacting freely. Maufort claims that in Grace-Smith's work "[the] verisimilar representations of ordinary life [are] interspersed with sudden intrusions of the supernatural world of ancestors, or indeed allusions to the universe of Māori mythology" (248). This idiom can be seen in *The Strength of Water*, written by Grace-Smith and directed by Armagan Ballantyne. Grace-Smith considered the Hokianga landscape where the film was shot—land, sea, cliffs—as characters in *The Strength of Water* ("NZ on Screen" website). Elements of the ecosystem interpenetrate the characters' lives and action. The film sets the impermanence of human life against the perceived permanence of land, reflecting the Māori whakataukī (proverbial saying), "whatungarongaro he tangata, toitū he whenua" [humans disappear but the land is permanent]. The film explores the evanescent borders between life and death. There exists a mauri in all things, even those which do not seem "alive" in the strictest sense.

The film's opening image is that of jellyfish, ten-year-old Melody's kaitiaki (guardians), swimming in a near-silent ocean. Bursting through the ocean's surface, the camera becomes a primal sea-creature moving into the air and suggests (re-)birth and renewal that are important thematic elements in the film. That initial focus on the ocean calls forth not the "unrestricted placelessness" that J. Douglas Porteous condemns as the "landlubber's viewpoint" but Tangaroa, the deity associated with the ocean and—in the creation narratives of some iwi (tribal groups)—Papa-tūā-nuku's first consort (89). The tides are associated with Hine-te-iwaiwa, the deity principally associated with weaving (Harrison, Te Kanawa and Higgins 124). The film also depicts the weaving together of people and its consequential restoration of spirit. Water also holds the power to cleanse in the Māori world, and when something tapu or restricted takes place, the tapu is often cleared or cleansed by water. In the Māori world, the land also holds its own mauri. In *The Strength of Water*, humans are initially depicted alongside—yet dwarfed by—the landscape. Early in the film, the viewer sees the lone figure of newcomer Tai walking along a deserted road, followed by a wide shot of the ten-year-old twins Kimi and Melody on a hill. They, too, are depicted as similarly tiny and insignificant against the landscape's vastness.

Elemental forces beyond human explanation and the porous margin between life and death form a core focus in the film. Its central narrative charts Kimi's journey of acceptance after Melody's accidental death while trapped in a junkyard freezer; it also examines the human interpenetration with the world of nature. The symbiosis between Kimi and Melody after Melody's death reflects the Māori concept of wairua—a "quintessence… spirit of a person which exists beyond death" (Te Whanake). Kimi holds Melody's asthma inhaler, and so he holds her life; she dies in the disused freezer because Kimi is not on hand with her medication. The junkyard in which Melody dies is filled with rusted car bodies, old machinery, plastics and abandoned whiteware, all haphazardly cast aside. However, these objects hold their own mauri. Turner poses the question "[what] if everything that happens is natural, including Love Canal and Alamogordo?" (41). Given Ka'ai and Higgins' definition (18), these items, too, have their own mauri. They are made from elements that are part of the planet's ecosystem and now lie abandoned, a testament to the excesses of human consumption and waste.

Kimi is drawn to his bedroom window as Melody dies in the freezer. Outside, he sees the ruru (morepork, owl), which in the Māori world are "linked with tapu (spiritual restriction), guardianship, forewarning, grief and awareness" (Te Ara "Ngā Manu: The Birds") and rarely seen during daylight. Similarly, moments after Melody dies, the rain—an "expression of grief" sometimes compared to tears falling during mourning in the Māori world (Te Ara "Tāwhirimātea: The Weather")—falls on the junkyard, and continues to fall during Melody's tangi (funerary ceremony). In the Māori world, tapu can be cleansed by water (Te Ara "Ngā Karakia"). In addition to tears being made of water, Armagan Ballantyne points out humans are primarily made of water ("The Strength of Water"), so once more the blurry boundaries between human and nature become evident. Moreover, the film suggests nature consists of more than that for which our eyes provide evidence. Even a psychological reading infers the presence of that which is not perceptible. The film depicts Te Ao Wairua (the Spirit World) as a force available to human understanding but potentially imperceptible through the five senses. It is important to reinforce the fact that these images appear not simply as visual metaphors in the film's mise-en-scène but are construed as palpable forces in the ecosystem inseparable from the characters' lives.

When Melody returns after her death, only Kimi can see her. Harriet Wild suggests that "[i]n his mourning, Kimi is not able to abandon his sister—or is it that Melody is not able to abandon Kimi?" (15). Wild concludes that Melody's presence after her death signals that Kimi, unable to accept his twin's death, creatively hallucinates her continuing proximity. Wild also notes that the distinction "between reality and the spirit world has been dissolved—yet reinstated elsewhere" as Melody is visible only to Kimi (15). Having said that, other characters tacitly acknowledge Melody's presence. Gibby, the children's father, comments on Kimi's use of "we" when ostensibly referring only to himself, and local teenager Tirea tells Kimi that "the whole town has eye trouble" when Kimi confides in her that no one but him can see Melody. Gibby's prescience of the nature of Kimi's outbursts is further signalled when Kimi awakes from a bad dream having wet his bed. While Joy, the children's mother, concerns herself with the pragmatic details of caring for her son and changing his sheets, Gibby's facial expression suggests that he accedes to the possible origin of Kimi's erratic behaviour. The dog, belonging to the twins' elder brother Gene, is ostensibly dead, but crawls into the Tapu House and Tai restores it to life. Its presence unnerves Gene, further reinforcing for him the evanescent borders between the living and the dead, and adding to his sense of guilt; the dog lived whereas Melody died.

Melody's presence remains palpable for Kimi but increasingly her wairua (spirit) slowly diffuses. Soon after her return, she forgets the words to the waiata she always sings. Later, she tells Kimi that she cannot feel anything and has no tactile—and Wild suggests, emotional (16)—awareness. She believes that she consists only of "skin and a bit of hair; there's nothing inside. My guts is gone" (*The Strength of Water*). Signalling that her wairua wants to be set free, she shows Kimi what it is like to be dead, but not free to return to Hawaiiki (the ancestral homeland of the Māori), in his dream of moving through deep water. Whilst, on one level, the dream duplicates the film's opening image of jellyfish gliding through the tranquil ocean, the image also acts as its terrifying doppelganger: the ocean appears dark as Melody flails, seemingly unable to manoeuver. A faint karanga (call of welcome) is discernable, calling Melody to a safe homeland she cannot reach, and we hear Melody's terrifying struggle for breath—or is it Kimi's?—underwater. Later in the film, Melody advises Kimi how to look after Aroha the chicken, effectively delivering her own poroporoaki—literally, her instructions before departure. Thereafter, she slips away, appearing with Kimi less frequently; we see her sleeping before Kimi's mission to cut Tirea's hair. Melody's intercession at this point further signals her imminent leaving; she tells him, "You can't keep her, Kimi. You don't do that to someone you love" (*The Strength of Water*). Aroha's death at the hands of the "Chicken Saver" symbolically cements Melody's impending departure. Kimi is drawn to the junkyard and finally to the freezer where Melody died. He approaches the freezer as he was not able to approach Melody's coffin at the tangi. The gecko lizard he sees on the freezer's lid is regarded as an omen, a "representative of Whiro, the god of darkness, evil and death" in the Māori world. When green geckos "lifted their heads and chattered"—as was the case at this point of the film—"it was seen as a bad omen" (Te Ara "Ngārara-Reptiles"). Conversely, lizards are also considered guardians whose function it was to watch over the dead (Te Ara "Ngārara-Reptiles").

Tai finds solace on the margin of the sea, betwixt and between land and ocean, which reflects his own liminality and dislocation. He takes up residence in his late grandfather's house—the Tapu House—and, in doing so, begins to find a sense of permanence of his own, although he regards himself as "bad luck" (*The Strength of Water*). Tirea tells him that his grandfather "looked after the mountain" (*The Strength of Water*), suggesting not only the symbiosis of land and humans, but also the immutable endurance that Tai so desperately needs. When Tai first enters the Tapu House, we see a spider crawling up the veranda post. The compound elements of the spider's name (punga—to anchor; werewere—a barnacle ["Ngata Dictionary"]) suggest its nature, and furthermore symbolise the fact that this place will become an anchor point for Tai. Like the evasive spider, Tai has dropped from sight when danger is imminent, and holds on by a thread. Nevertheless, the Tapu House and the area in general will enfold him in its web, interrupting his flight. This is the same veranda post that Tirea leans against when she first visits Tai at the Tapu House, thus associating her with that spider, and signalling the delicate, near-invisible string of affinity that has begun to be woven between her and Tai. Later, when walking through the labyrinthine tangle of trees in the bush, they can speak frankly. Tai tells Tirea about a trip to the top of the Sky Tower in Auckland: "The floor at the top was glass—can lie flat and look down, can see all the small buildings, tiny people running around doing their shit. If I look close enough I might see me down there" (*The Strength of Water*). A wide shot of hazy, majestic hills and the ocean at dawn follows this exchange, thus reinforcing that very sense of human insignificance that Tai experienced on the Sky Tower. The tower's transparent floor, like Melody's jellyfish kaitiaki, appears insubstantial but provides a protective aegis. Water also protects Tai when Gene excludes him from Melody's tangi. Tai toils back to the Tapu House through the chest-deep sea, with the ocean acting as an agent to help cleanse his tapu associated with death, "loosing [him] from what is destructive" (Shirres 67). His guilt can be assuaged more so than Gene's, and his healing can slowly begin.

Tirea's disposition intertwines itself with the land. She tells Tai on the hilltop, "I've got that feeling—you know—when you just want to jump off." This moment deals not simply with the vertiginous sensation of wanting to jump off a high object, but also with Tirea's quest to find who she is and to begin to spread her wings, to rise up and to fly. Tirea is often seen on horseback, flying as best she can from her landlocked position. Tirea's very name reinforces this idea; it refers to the second night of the new moon, when the moon's radiance is just beginning to be seen. So it is with Tirea; she is a fledgling. This name can also be translated as "a flash; a radiation" or as the moon's ascent from the underworld (Roberts, Weko and Clarke 7), all of which features can be seen in Tirea. Kimi's feelings about the loss of Melody become entangled with his crush on Tirea, and his resentment over her relationship with Tai. Cutting Tirea's hair is his way of robbing her of the powers he perceives her to hold as a Patupaiarehe, one of the fairy folk who could "lure people away from safety with the music of their flutes, and had magical powers and special knowledge." There is a belief that "red-haired Māori are their descendants" (Te Ara "Understanding the Natural World"). Kimi atones for cutting Tirea's hair by returning a clump of it to her, and his journey out of grief continues as Tai relates to him Melody's last moments in the freezer. Kimi asks, "Did she cry? Did it hurt?" to which Tai can only reply "She just ran out of air" (*The Strength of Water*). Kimi, too, has become exhausted by his attempts to hold onto Melody. He finally says goodbye to her, on the cliff-top above the sea, embracing her before her spirit is released over the ocean to Hawaiiki. Kimi hears the waiata Melody sang, an incantation for the rains to cease, and the viewer knows she now dwells in the care of her guardians, the jellyfish. Kimi can replace Melody's photo on the picture wall for the dead at the marae (traditional meeting space), re-locating her in Te Ao Wairua. He walks home barefoot, cleansing himself in the stream, finally removing his tapu after his association with the dead. The film's depiction of human and natural forces positions these elements as relational, wherein humans interconnect with their kaitiaki and with the landscape, rather than anthropocentrically where humans only consociate to other humans as unchallenged masters of the universe.

The Collision of Worlds: *When Sun and Moon Collide*

Although most of *When Sun and Moon Collide* takes place in Isaac's tearooms, its situation "between the mountains, under a vast sky" means the landscape's life-force breathes beneath the play's action (Grace-Smith xv). In her "Author's Preface," Briar Grace-Smith recalls her inspiration for writing the play: a long daily car trip in which she was passenger, the frequent sight of an abandoned tearooms [sic] and a thin girl running through fields "as if she were being chased." These human elements connect inexorably to the landscape as Grace-Smith details: "...all of us were bound together by huge brooding skies and the dark ridges of the Tararua mountains" (xii).

At the play's opening, Declan Kopu, recently released from prison, returns to his life-long friend Isaac's tearooms to seek Isaac's help. The phases of the moon, especially the umbrageous Mutuwhenua (dark phase of the moon) have caused Declan's alter-ego "Jason" to emerge and perhaps wreak havoc. During Mutuwhenua, conventional behaviour would seem to be abrogated, and the delineation between order and chaos becomes slippery.

Much of the play's action occurs in Isaac's mostly unpatronised tearooms. His former customers now frequent Café Astounding across the road and Isaac appears to make little—if any—money, yet his tearooms provide a sanctuary and a sense of place for the play's characters. Nothing changes in the tearooms, not even the menu, but a sense of foreboding overshadows the action, like the mountains that overlook the tearooms. Isaac says: "[i]t's not safe. New Zealand is extremely dangerous. It isn't some island paradise..." (29). Like the land expropriators of the nineteenth century who—in the language of conquest—violated the natural world, and promoted a monoculture of people and environment, land speculator Vic Dwyer not only sells land, but also has instruments "[f]or hurting" people (66). Francie—Vic's anorexic ex-partner—has been his victim and holds a locket as proof that he murdered the missing Danish backpackers. Francie finally discloses to Declan the depth of Vic's malignant behaviour. The urgency of her need to stem Vic's evil has a correspondence in the natural word. She tells Declan, "[i]t's all falling away from under us. Even the earth's shifting" (67). Declan responds with an alternative reading of their environment through observation of the sky:

> The sun is out with the moon. Korekore te Rawea. Twenty-third phase of the moon. Not a fruitful night. Food is scarce… But await, my dear, the turn of the tide. The tide will turn. The tide will turn. (67)

The proximity of sun and moon—the play's title—signals the beginning of a shift in Francie's and Declan's circumstances and the righting of unpunished wrongs.

The personification of natural elements, too, shows interpenetration between human and natural worlds. Declan, manifesting the persona of his Koro (grandfather), tells his younger Declan-self: "The river might be shallow, but she's mighty pissed off tonight" (75). Later, police officer Travis recalls her mother's whakataukī, which applies elements from the natural world to human experience, "[t]he kowhai[5] would never've met the tūī[6] had it not found its courage to bloom," thus illustrating a human world amplified by and saturated with natural forces, not disconnected from them (69). Francie tells Isaac, "[w]hen you watch the world through glass, Isaac, it makes it real easy to believe that it's not real" (81). Indeed, Francie's statement calls to mind Grace-Smith's description of herself watching the world through a car window on long trips from Paekākāriki to Palmerston North and composing the play's narrative from disparate and unrelated elements, unifying them through the natural features of sky and landscape ("Author's Preface" xii).

The forces of regeneration, re-birth and turning tides permeate the human world, yet the stasis Declan yearns for proves impossible. Francie tells him, "[n]othing's still. Not even the planets. The earth circles the sun, and the moon circles the earth. We're all stalked or being stalked by something" (85). Isaac's decision to leave at the end of the play represents a decision to encounter the world, to "hitch a ride. See where I end up" (83).

Francie begins the play "never notic[ing] the sky or the sea," signalling her inability to fully inhabit and function in the surrounding ecosystem (Grace-Smith 64). She has stopped eating, a futile attempt to control her world. What she fails to understand is that she is connected to the entire ecosystem, and that her own sense of control must be

---

5 *Sophora tetraptera*; native New Zealand tree with yellow flowers that attracts birds, especially nectar-eaters such as the tūī.

6 *Prosthemadura novaeseelandiae*; native New Zealand nectar-eating bird.

adapted to its exigencies through the Māori concept of utu a sense of balancing—until "[any] matter is considered *ea* (balanced)" (Ka'ai *et al.* 18). As contemporary ecologists might point out, rather than exhibiting "wholeness or stability," the biosphere is constituted by "nonlinear, discontinuous order" that corresponds with the notion of utu. A tapu is cleansed through dialogue between humans and the natural world (Howarth 75).

Te Mutunga: Conclusion

An accurate reading of the three texts studied arrives through a cognisance of the world as a mosaic of human and natural sentience. Nature emerges not as a metaphor for human experience but as sibling to humanity and culture. However, to consider the texts in this way requires a re-evaluation of a contemporary western worldview "with its faith in reason, intellect and progress," a worldview that has "created an immense realm of silences, a world of "not saids" called nature, obscured in global claims of eternal truths about human difference, rationality, and transcendence" (Manes 17). The three texts examined in this essay render non-human "voices" palpable, with each voice expressing its own mauri or life-force. Humans and other elements in the landscape share a genealogical connection to primal parents Papa-tūā-nuku and Ranginui, and beyond, resulting in characters being changed or modified by forces in the landscape. Transformation, however, does not necessarily equate "progress" in the scientific sense, nor does it suggest a "taming" of nature. Instead, it leads to a sense of concordance between humans and the natural world. The mountains, sky, animals and birds are represented as forces of transformation, cleansing the tapu of death, war and loss.

"Whatungarongaro te tangata, toitū te whenua"
["People disappear, but land is permanent"]

Works Cited

Byrd, Jodi A. "(Post) Colonial Plainsongs: Towards Native Literary Worldlings." Ed. Don Trent Jacobs. *Unlearning the Language of Conquest: Scholars Expose Anti-Indianism in America.* Austin: U of Texas P, 2006.

Cameron, Sharon. *Writing Nature: Henry Thoreau's Journal*. Oxford and New York: Oxford UP, 1985.

Duerr, Hans Peter. Trans. Felicitas Goodman. *Dreamtime: Concerning the Boundary Between Wilderness and Civilization*. Oxford and New York: Basil Blackwell Ltd., 1985.

Grace-Smith, Briar. *The Strength of Water*. Dir. Armagan Ballantyne. Pandora Film, 2009.

Grace-Smith, Briar. *When Sun and Moon Collide*. Wellington: Huia Publishers, 2007.

Harrison, Pakaariki, Kahu Te Kanawa and Rawinia Higgins. "Ngā Mahi Toi: The Arts." Ed. Tānia M. Ka'ai, John C. Moorfield, Michael P. J. Reilly and Sharon Mosley. *Ki Te Whaiao: An Introduction to Māori Culture and Society*. Auckland: Pearson Education New Zealand, 2004. 116-32.

Howarth, William. "Some Principles of Ecocriticism." Ed. Cheryll Glotfelty and Harold Fromm. *The Ecocriticism Reader: Landmarks in Literary Ecology*. Athens and London: The U of Georgia P, 1996. 69-91.

Ihimaera, Witi. *Woman Far Walking*. Wellington: Huia Publishers, 2000.

Ka'ai, Tānia M. and Rawinia Higgins. "Te Ao Māori: Māori World-View." Ed. Tānia M. Ka'ai, John C. Moorfield, Michael P. J. Reilly and Sharon Mosley. *Ki Te Whaiao: An Introduction to Māori Culture and Society*. Auckland: Pearson Education New Zealand, 2004. 13-25.

Klein, Ulrich. "Belief-Views on Nature—Western Environmental Ethics and Maori World View." *New Zealand Journal of Environmental Law* 4 (2000): 81-120.

Learning Media: Te Pou Taki Kōrero. "Ngata Dictionary." Web. 10 December 2012. http://www.learningmedia.co.nz/ngata/.

Love, Glen A. "Revaluing Nature: Toward an Ecological Criticism." Ed. Cheryll Glotfelty and Harold Fromm. *The Ecocriticism Reader: Landmarks in Literary Ecology*. Athens and London: The U of Georgia P, 1996. 225-40.

Manes, Christopher. "Nature and Silence." Ed. Cheryll Glotfelty and Harold Fromm. *The Ecocriticism Reader: Landmarks in Literary Ecology*. Athens and London: The U of Georgia P, 1996, 15-29.

Mitchell, John Hanson. *Ceremonial Time: Fifteen Thousand Years on One Square Mile*. New York: Anchor Press/Doubleday, 1984.

Maufort, Marc. "Recapturing Māori Spirituality: Briar Grace-Smith's Magic Realist Stage Aesthetic." Ed. Maufort, Marc and David O'Donnell. *Performing Aotearoa: New Zealand Theatre and Drama in an Age of Transition*. Brussels: P.I.E. Peter Lang, 2007, 247-67.

NZ on Screen "The Strength of Water." Web. 7 March 2012. http://www.nzonscreen.com/title/the-strength-of-water-2009.

Porteous, J. Douglas. *Landscapes of the Mind: Worlds of Sense and Metaphor*. Toronto, Buffalo, London: U of Toronto P, 1990.

Roberts, Mere, Frank Weko and Liliana Clarke. "Maramataka: The Maori Moon Calendar." Research Report, Lincoln University. Web. 25 November 2012. http://www.lincoln.ac.nz/Documents/2333_RR283_s6506.pdf.

Seed, John, Joanna Macy, Pat Fleming, and Arne Naess. *Thinking Like a Mountain: Towards a Council of All Beings*. Gabriola Island, BC: New Society Publishers, 1988.

Shirres, Michael P. OP. *Te Tangata: The Human Person*. Auckland: Accent Publishing, 1997.

Tallmadge, John and Henry Harrington, eds. *Reading Under the Sign of Nature: New Essays in Ecocriticism*. Salt Lake City, Utah: U of Utah P, 2000.

Te Ara: The Encyclopedia of New Zealand On-line. "Ngā Karakia a te Māori: Traditional Māori Religion." Web. 11 December 2012. http://www.teara.govt.nz/en/nga-karakia-a-te-maori-traditional-maori-religion/5.

Te Ara: The Encyclopedia of New Zealand On-line. "Ngā Manu: The Birds." Web. 23 November 2012. http://www.teara.govt.nz/en/nga-manu-birds/3.

Te Ara: The Encyclopedia of New Zealand On-line. "Ngārara-Reptiles." Web. 25 November 2012. http://www.teara.govt.nz/en/ngarara-reptiles/1/2.

Te Ara: The Encyclopedia of New Zealand On-line. "Tāwhirimātea: The Weather." Web. 23 November 2012. http://www.teara.govt.nz/en/tawhirimatea-the-weather/4.

Te Ara: The Encyclopedia of New Zealand On-line. "Understanding the Natural World." Web. 25 November 2012. http://www.teara.govt.nz/en/understanding-the-natural-world.

Te Whanake: Māori Language On-Line. "Māori Dictionary." Web. 25 November 2012. http://www.maoridictionary.co.nz.

Turner, Frederick. "Cultivating the American Garden." Ed. Cheryll Glotfelty and Harold Fromm. *The Ecocriticism Reader: Landmarks in Literary Ecology*. Athens and London: The U of Georgia P, 1996. 40-51.

White, Jr., Lynn. "The Historical Roots of Our Ecological Crisis." Ed. Cheryll Glotfelty and Harold Fromm. *The Ecocriticism Reader: Landmarks in Literary Ecology*. Athens and London: The U of Georgia P, 1996. 3-14.

Wild, Harriet. "Creativity and Mourning in *The Strength of Water*." *Illusions* 43.44 (Spring 2012): 13-18.

Performing the Spirit of the Earth

Multi-faceted Aesthetics of Ecology in Contemporary Indigenous Drama[1]

Marc Maufort

Université Libre de Bruxelles (ULB)

Contemporary Indigenous drama in North America and Oceania offers a unique comparative vantage point from which to study the multiple configurations of what could be called the "aesthetics of ecology." As Lawrence Buell reminds us, ecocritics generally distinguish between the notions of "space" and "place." While "space" designates an abstract concept, "place" is endowed with emotional and spiritual features, thus pointing to Indigenous peoples' sense of belonging to a specific environment (Buell 63). "Place" corresponds to the kind of ecology that Huggan and Tiffin describe as "a sphere of co-dependent interaction that connects people to other ecological beings, both animate and not, that share their phenomenal life-world" (70). This contrasts sharply with the Western view, upheld by the Enlightenment, of nature as a space of economic exploitation (Clark 7). "Place" thus suggests the erasure of any rigid distinction between Nature and Culture, emphasizing the web of connections existing between the human and the non-human. The thematic implications of "place" in Indigenous drama have been studied by a number of critics. As early as in 1996, Gilbert and Tompkins recognized the prominence of the land in postcolonial drama (cf. 156). More recently, Christy Stanlake has examined the complex ways in which the importance of the land makes it an almost central character in Indigenous North American drama (Stanlake 41-43).[2] However, few

1 I wish to express my gratitude to Birgit Däwes and David O'Donnell who commented on earlier drafts of this essay and helped strengthen its argument. I am also deeply indebted to my colleague Franca Bellarsi for generously sharing with me her infectious enthusiasm for ecocriticism.

2 While acknowledging this Native vision of the land, we should nonetheless guard ourselves against what Greg Garrard calls the myth of the "ecological Indian," which wrongly suggests an unproblematic relationship between Natives and Nature (Garrard 129).

critics hitherto have analyzed in detail the wealth of the theatrical aesthetics Indigenous playwrights have devised in order to articulate this sense of place. When tackling this aspect of Indigenous dramaturgy, I would contend it is essential to bear in mind its particular relationship to the contemporary phenomenon of globalization. Late-twentieth century globalization contributed to the dissemination of knowledge about Indigenous literary production around the world, as Native plays increasingly made it into print and professional cooperations between Indigenous theatre companies from different countries developed significantly. In addition, Indigenous playwrights saw their works showcased at international festivals.[3] However, as Buell points out, globalization also tends to promote a discourse of homogenization, which can hamper the development of a localized sense of place (Buell 92). As Huggan and Tiffin put it, Indigenous playwrights are forced to negotiate the tension between "embeddedness (locality) and interconnectedness (globality)" (Huggan and Tiffin 69). They must counter the commercial impulse typical of late capitalism and globalization to fetishize the local, i.e. to ascribe to it rigidly defined and stereotypical values (Huggan and Tiffin 63). This essay therefore seeks to shed light on the highly unique aesthetics of ecology articulated by Indigenous playwrights in order to reject the homogenizing mould of globalization. To that end, this contribution places in conversation plays from Canada, Australia, and New Zealand, in an effort to highlight the diverse vocabularies used to enact the Indigenous bond with the spirit of the Earth.

I

Tomson Highway's *Ernestine Shuswap Gets Her Trout* (2004) offers a historiographic approach to ecojustice, while foregrounding female agency.[4] The play recounts the 1910 visit of Wilfrid Laurier, then-Prime Minister of Canada, to the Thompson River valley in British Columbia. On this occasion, he met with Native chiefs whose fishing rights had been confiscated. As the play draws to a close, the heroine, Ernestine

[3] An instance of such international collaborations can be found in the work of Māori playwright Miria George. Her 2009 play, *Urban Hymns*, was developed in collaboration with the Toronto-based Native Earth Performing Arts company (George 121). Further, Highway's *The Rez Sisters*, to give only one significant example, was programmed at the 1988 Edinburgh Festival and at the 1993 New York Theatre Workshop's New Directors/New Directions series (Gilbert 393).

[4] *Ernestine Shuswap Gets Her Trout* has not failed to elicit critical attention. It is discussed in Huggan and Tiffin (124-25), in Stanlake (27, 149-52), in Nelson Gray's "Bringing Blood to Ghosts" (Gray 28-29), and in De Wagter's study of multi-ethnic North American drama (De Wagter 183-84, 189-200). None of these studies, however, examines Highway's work within the larger context of globalization. In this essay, I use the concept of "ecojustice" in much the same way as Timothy Clark, i.e. as the study of the link between environment and social justice issues (Clark 120).

Shuswap, confronts the authorities far more boldy than her inefficient husband. The ecological dimension of the play is asserted at the outset, as Highway stresses the importance of a specific feature of the set: "…a plain white backdrop hangs slashed by a line, curved, horizontal: the land in silhouette" (10). The prominence of the earth is again foregrounded in the prologue, in which we hear "…the gurgle of a river—rich, evocative, the voice of a land" (13). The use of language is equally crucial as the representation of the land in this play. As Christy Stanlake has indicated, "language stems from the earth" in the Native world view (Stanlake 87). Although he belongs to a Cree cultural background, Highway explores the dynamics of Shuswap language:

> *The language spoken by the women in this play, it must be stressed, is* not *English.* […] *In this play, they speak Shuswap, a tongue that works according to principles, and impulses, different entirely from those that underlie, that "motor", the English language. For instance, because the principle that "motors" the Shuswap language is, in essence, a "laughing deity" (i.e. the Trickster), it is hysterical, comic to the point where its "spill-over" into horrifying tragedy is a thing quite normal, utterly organic. That is to say, as in most languages of North America (that I know of anyway), the "laughing god" becomes a "crying god" becomes a "laughing god," all in one swift impulse*. (11)

Highway's observations suggest an extremely fluid, ever-shifting vision of the world, one in which land, river and language are linked as expressions of a Native deity that can assume different shapes. This elusive tricksterish characteristic typifies the entire play, in an echo of Highway's classic *The Rez Sisters*.[5] Thus, the relationship between humans and the non-human world is articulated through tricksterish aesthetics.

As the following analysis will demonstrate, in *Ernestine Shuswap Gets Her Trout*, Highway tweaks the codes of Aristotelian linearity. He achieves this precisely through his tricksterish mode of writing, i.e., the combination of such features as storytelling, the erasure between comedy and tragedy, the use of a Western classical music soundtrack, and the extension of the naturalistic template towards a magic realist idiom. The flexibility of this tricksterish aesthetics enables him, as early as in the prologue, to articulate the profound ecological link between Natives and the earth. Given the all-female cast of this play, the land is markedly gendered feminine. As a character named Annabelle claims, the Thompson

---

5 Although the use of English in this play is meant to reproduce the elasticity of the Shuswap world vision, Highway does not insert any passage in the Shuswap language. In his earlier *The Rez Sisters*, however, the playwright subtly mixed English and Native language, a sign of his plurilingual approach to dramaturgy. For more information about *The Rez Sisters*, the reader could turn to my book, *Transgressive Itineraries* (149-52).

river cannot be distinguished from human flesh: "That winding, greenish-bluish river that's so busy just a-weaving its way through the mountains [...] through our windows, through our doors, through our children, through our lives, through our dreams, our hearts, our flesh, our veins, our blood" (16). Throughout the play, Highway underlines the osmosis existing between land, river, animals, language and Native people. When asserting the historical importance of this meeting, Isabel also implicitly compares the Thompson river to a human body: "I can feel it deep inside my hot Shuswap Indian capillaries coursing up and down like the great Thompson river as it comes cascading from the Lollipop Mountains away over yonder" (27). The metaphor is prolonged to incorporate animals: "I, Isabel Thompson of Kamloops, B.C., was born with eyes of a large, brown owl, see?" (30). As Ernestine is picking saskatoons in preparation for the banquet, the stage directions suggest she literally becomes a part of the vegetation: "[...] *the shimmer of leaf, branch, and stalk doing shadow-play across ERNESTINE SHUSWAP's handsome face, her body*" (39). Further, Highway likens the feverishness of food-making to the hysterical and farcical features of the Shuswap language. Ernestine feels extremely nervous when preparing the trout stuffing: "[...] the Great Big Kahoona of Canada, who's as French as a cheese and Catholic as a rosary, will be eating at that big banquet, right beside our humble little Indian chief, tonight, Thursday, the twenty-fifth of August, 1910 at nine P.M. So you see, it all makes disgustingly delightful, monstrously delicious, juicy-Lucy kind of sense, don't you think?" (42). Isabel continues in this tricksterish vein when alluding to Barnabas's linguistic abilities, her long list bordering on farcical exaggeration:

> [...] Barnabas, who can speak not only his Native Thompson, but Shuswap, Okanagan, Haida, Tsimshian, Tlingit, Klingon, Haisla, Yakima, Santee Sioux, Oglala Sioux, Lakota Sioux, Chalala Sioux, Dogrib, Yellowknife, Blackfoot, Cree, Arapaho, Navaho, Assiniboine, Winnebago, Sauk and Fox, Seminole, Wichita, Choctaw, Cherokee, Chipewyan, Chippewa, Chickasaw, Chickadee, Kickapoo, Boogaboo, Micmac-paddy-walk-give-a-dog-a-bone *and* ten words of English? (51)

The conclusion of Act I ultimately reasserts the link between language and the land. While Ernestine and Isabel are picking saskatoons, "[...] *the land is talking, through them*" (56). In a hybrid conflation of the Native world vision and Western culture, the sound of classical music merges with the language of Nature: "[...] *the cello music rises, fades and dies... as behind it (or under it), does the river*" (56). As the sense of oppression increases in Act II, the female protagonists start neighing in order to protest against the cancellation of the Shuswap language by the colonial authorities (68). Enestine even seeks the approval of the audience about the quality of her neighing, in a metatheatrical moment replete with comic overtones (75). Isabel, who has come to resemble a racoon, starts mooing to express her unhappiness (72-73). As Delilah Rose dies, her neighing resonates with both farcical and tragic connotations: "[…] *she neighs. And neighs and neighs and neighs, galloping up and down the stage like a horse. Until, gradually, human language starts 'bleeding' from her neighing; it is as if she were speaking in tongues*" (81). Towards the end of the play, the text of the Laurier memorial, recording the Natives' complaints about the confiscation of their fishing rights, seems to emanate from Nature: "[…] *the speech mumurs on—in effect, it is the river talking* […] *Its shuswap translation, moreover, 'circles' the auditorium like a restless breeze*" (86). For the first time, Shuswap language is supposed to be heard in the auditorium, surprisingly uttered by a male voice. The happy climax of the banquet seems to suggest ongoing Native defiance. As Daisy the cow moos happily, "[…] *the gurgle of a river surfaces, balloons, and fades, rich, evocative, the voice of a land, and blackout*" (90). Thus, the play concludes on an affirmation of the power of Nature, a stance recalling the first moments of the work. In a further hybrid fusion of Western and Native cultures, the sound of Beethoven's Sonata for Cello op. 69 optimistically "*sounds like a river, a river flowing smooth, majestic, dignified, and powerful*" (90).

When one considers other literary devices used in this work, one cannot help noticing a similar equation between human beings and Nature. In *Ernestine Shuswap Gets Her Trout*, storytelling and magic realism combine to express the tragedy inherent in the lives of the three female protagonists, Annabelle, Delilah Rose, and Ernestine. Delilah Rose progressively reveals that her lover, Billy Johnson, is the son of the Protestant leader who confiscated Native fishing rights. This puts her in a very awkward position towards the members of her Catholic community, who regard her as as a traitor to her culture. At first, she explains how her love was deeply connected to the Thompson river: "That's what made me fall in love with you, Billy Boy Johnson […] that night among the rushes on the banks of the Thompson river, when you asked if you could kiss me on…" (48). She also reveals that she is now pregnant with Billy Johnson's child, which betrays her predicament of hybridity (49). In the last moments of Act I, Delilah appears as a goddess emerging from the river, the presence of which is suggested through the tablecoth she is stitching. She is "[…] *looking like a goddess emerging from a river (the length of white muslin)* […]" (56). Further, the tablecloth, we are told more explicitly, "*looks like a river*" (63), a reference to Highway's transformational tricksterish aesthetics. In a gesture of despair, Delilah Rose kills her baby before committing suicide. In a magic realist moment suggesting the blurring of any distinction between the real and the supernatural, she describes her ghost as hovering over the Thompson river.[6] Her eyes have now joined those of the ghosts of two previously murdered men. In this particular passage, magic realism reflects a Native vision of the world that denies any ontological difference between ordinary reality and the world of spirits: "DELILAH ROSE: [...] To this day, a third pair of eyes hang high over the river in the mist, on nights when the moon is out, a third pair of eyes hang high over the river in the mist, like lamps. [...] this woman's head in the mist [...] looks out at a banquet that sits on her veil of pure white muslin, a banquet that sits on her beautiful veil, a veil they call the Thompson River" (86). In this play, magic realism is thus deeply linked with the features of the natural environment.

As she tells the tragic story of her own life, Annabelle also points out the link between her predicament and ecojustice: "A white-skinned cowboy named Oliver Clapperton killed Johnny Okanagan, my dear late husband. [...] Killed him for a piece of land, *his* land. Johnny Okanagan's little piece of trap-line, yes. And I forgive him, Delilah Rose Johnson" (84). Ernestine's own storytelling leads to a final sense of forgiveness and healing. She first tells the tragic story of the murder of a Native, Franklin Coyote, who was killed by white settler Don Jenkins in retaliation for stealing his chickens. In a gesture of further revenge, Franklin Coyote's brothers killed Don Jenkins at Indian Point, near the Thompson River. In a mood of magic realism, Ernestine's story then suggests that the ghosts of the two murdered men keep haunting the Thompson River area: "And the blood of Don Jenkins Sr. and the blood of Franklin Coyote still hang in the air, like a mist, over the waters of the Thompson River. And they say that inside that mist,

[6] The theoretical work of the European critic Jeanne Delbaere sheds considerable light on the all too often elusive notion of magic realism, most notably her essay "Psychic Realism, Mythic Realism, Grotesque Realism: Variations on Magic Realism in Contemporary Literature in English." Magic realism, Delbaere argues, is often used by postcolonial or minority writers to express a reaction against the center, against hegemonic society. It presupposes the conflation of concepts or objects one would not normally associate within a Western rational context, thus asserting the validity of an Indigenous worldview in which ordinary reality and the supernatural are contiguous rather than antithetical. In what Delbaere terms "psychic realism," a variant on magic realism, the "magic" constitutes "almost always a reification of the hero's inner conflict" (251). This form of magic realism records the character's fissured vision of the real. Delbaere's second subcategory defines a form of "grotesque realism," which tends to further distort and amplify reality. Delbaere suggests that "grotesque realism" be used "for any sort of hyperbolic distortion that creates a sense of strangeness through the confusion or interpenetration of different realms like animate/inanimate or human/animal" (256). A third subcategory, which Delbaere terms "mythic realism," emanates from the supernatural features of the environment itself rather than from the character's psyche (252-53). In an ecocritical discussion of Indigenous drama, Delbaere's notion of "mythic realism" proves to be quite useful, particularly in the case of *Waiora*, precisely because it evokes a link between magic realism and the environment.

late at night, when the moon is out? You can see eyes in them, like lamps, two pairs of them, hanging there waiting and waiting and waiting…" (67-68). The subsequent stage directions amplify this magic realist fusion between ordinary reality and the supernatural: the spectators are shown these two pairs of eyes floating in the background. Later in the play, when Ernestine decides to catch a trout without the help of her useless husband, magic realism becomes the dominant mode. Ernestine is seen underwater, swimming like a fish, which erases the distinction between human and animal states of being. The stage atmosphere becomes increasingly surreal: "*And, suddenly, ERNESTINE is not only on the banks of the Thomson River, she's* in *it, walking on its bottom as in a dream ("river-sound" ebbs and flows throughout the scene). And as in a giant aquarium, fish "swim" by her* […]" (78). Isabel's subsequent reading of the Laurier Memorial reinforces this magic realist mood, as her voice sounds "*unreal, dream-like, otherworldly*" (78). Once again, magic realism is rooted in the perception of Nature. In the final scene of the play, Ernestine serves her trout at the banquet, a joyful ceremonial event that also looks like a "*funeral*" (87). The description of the trout reveals a surreal exaggeration evocative of magic realism, the more so as the ghost of Delilah Rose can be seen in the background: "*ERNESTINE* […] *comes with her trout on a platter, a great big trout, biggest one you've ever seen. Like a priestess in a ritual, she descends, slowly passes it to Isabel, who slowly passes it to Annabelle, who slowly passes it to the ghost of the very dead DELILAH ROSE JOHNSON* […]" (88). In addition, the scene resembles "The Last Supper, *with Laurier as Christ and the chiefs as the Apostles*" (88). The trout symbolically reinforces Highway's use of Christian symbolism, which bespeaks his hybrid vision of reality. The implied metaphor is that Nature itself becomes similar to the body of Christ and is to be ingested by the banquet participants in a sense of holy communion and regeneration. The play concludes on an assertion of Native culture: the voice of the land can be heard again.

All in all, in *Ernestine Sushwap gets Her Trout*, Tomson Highway astutely combines storytelling and magic realism in a fluid tricksterish dramaturgy asserting the instability of the human/Earth relationship. In this way, he offers an innovative enactment of Native ecology through a potent theater aesthetic.

II

Like Highway's play, Andrea James's *Yanagai! Yanagai!* (2003) deals with ecojustice issues.[7] Its displays a markedly historiographic approach: it is conceived as a protest against Judge Olney's 1998 Native Title

[7] As the program notes published with the play indicate, Andrea James "is a descendant of the Yorta Yorta nation, Kunai Aboriginal nation and Poland" (vii).

Judgment in the Federal Court of Australia, which denied Yorta Yorta peoples' claims to the land of their ancestors. This judgment was upheld by the High Court of Australia in 2002. As a gesture of rebellion, this play asserts the fundamental connection between Yorta Yorta people and their land. In an echo of Highway's tricksterish aesthetic, *Yanagai! Yanagai!* can be regarded as a re-enactment of the Yorta Yorta myth of creation, i. e. what is generally referred to in English as the Dreaming. As stories of origins, Australian Aboriginal Dreamings are specific to each Indigenous nation on the continent. As Jennifer Sabbioni reminds us, during the Dreamtime, "mythical beings emerged from the earth, the water, and the sky and assumed a variety of forms and identities. They roamed the vastness of a barren landscape, stopping and engaging in various activities through which they created mountains, rivers, water holes, flora and fauna, and the indigenous people, who were appointed as caretakers of the world that surrounded them" (Sabbioni xx). As Sabbioni concludes, "It can be said that the indigenous people of Australia were really the first environmentalists in the world" (Sabbioni xxi). Historian Russel Ward concurs when underlining the ecological interconnectedness between Aboriginal people and the land: "The tribe, its people and its territory belonged to each other so completely that Aboriginal Australians saw the three as one entity […]" (Ward 14). As Wally Caruana indicates, "The Dreaming focuses on the activities and epic deeds of the supernatural beings and creator ancestors such as Rainbow Serpents, the Lightning Men, the Wagilag Sisters, the Tingari and Wandjina, who, in both human and non-human form, travelled across the unshaped world, creating everything in it and laying down the laws of social and religious behaviour" (Caruana 10). Further, Caruana claims that the Dreaming lies at the source of Aboriginal Art, including rock and bark paintings (Caruana 11). One could therefore argue that *Yanagai! Yanagai!* is a dramatic equivalent of those Dreamings evoked through traditional Aboriginal art. Dreamings are deeply linked to specific places. This undoubtedly stresses the link between people and their local environment, thus countering both colonization and Western globalization: "The continent of Australia is covered by an intricate web of Dreamings. Some relate to a particular place or region and belong to those who reside there, other Dreamings travel over vast distances and connect those whose lands they cover" (Caruana 10). In addition to its inherently ecological dimension, the Dreaming also suggests a non-Western concept of time. According to Andrew Sayers:

> From the moment of conception an Aboriginal person has a connectedness to the realm of the spirit and specific Dreamings: through an individual's conception site he or she is connected in a particularly intimate way to a specific locality. […] One of the complexities of the concept of the Dreaming lies in its sense of time, different from the linear time that governed the lives

> of the colonists. An important characteristic of the Dreaming is its continuity: whilst it refers to the creative events of the past, the Dreaming is also alive in the present. (Sayers 18-19)

Yanagai! Yanagai! celebrates these features: a strong sense of embeddedness is paired with a visual representation of the Yorta Yorta Dreamings surviving in contemporary Australia. As the ensuing close reading of the play will attempt to demonstrate, Andrea James develops a highly innovative stage aesthetics that expresses this ecological sense of place through a subtle use of storytelling, court transcripts and magic realism.

In the introduction, Yorta Yorta Elder Wayne Atkinson tells the reader that "Yanagai" means "go away" in the Yorta Yorta language (n.pag.). As the title of the play, it echoes the historical resistance of Yorta Yorta people when confronted with the arrival of white settlers in the nineteeth century. From the outset, then, *Yanagai! Yanagai!* qualifies as protest drama. In the introductory statement, the playwright emphasizes her nation's traditional connection to land and culture, particularly the Murray river (Yorta Yorta territory is situated north of the State of Victoria). The importance of the river here recalls the prominence of the Thompson River in Highway's *Ernestine Shuswap Gets Her Trout*. Andrea James further describes her play as a "war cry" which "provides a forum for the Yorta Yorta language to be spoken into space again—to flourish and thrive—in a cultural ceremony written specifically to prove what Judge Olney could not or would not see—our traditional connection to the land" (n. pag.). *Yanagai! Yanagai!* indeed contains many passages, meant to be spoken in the Yorta Yorta language, for which translations are provided in the printed version of the play. This linguistic feature echoes Highway's distrust of the English language in *Ernestine*, although Highway came short of actually using Shuswap language. The description of the setting further reinforces the ecological dimension of the play, while introducing the notion of magic realism and storytelling as central aesthetic devices in the play: "Yanagai! Yanagai! is set in a mythical landscape on the banks of a mighty river. We call him Dhungula [The Murray River]. This river is made before our eyes. A sparkling celebration" (n.pag.) In addition, trees are seen floating in the space, while spirits appear in them, a magic realist conflation of ordinary reality and the supernatural. These trees are alive, Andrea James tells us. Animals fully belong to this universe: "In bright blue neon the word 'maniga' [to fish] floats in the air. […] A giant cod puppet floats magically in the space […]" (n.pag.). Further, Andrea James asserts the central role of storytelling as linked to the Dreaming: "Timezoned in the dreaming, present and future, a clan of storytellers have banded together to remember a beautiful place they once knew" (n.pag.). In the first scene, the storytellers eloquently express this lack of differentiation between human beings and Nature. Trees, equated with the blood of people, are described as the holders of spirits (1). The Murray River is compared to people's "life blood" (2). Stories thus celebrate the Yorta Yorta language (3) and culminate in the storytellers' assertion: "We are here" (4).

Throughout the play, storytelling merges with a form of magic realism emating from the land. This literary device is particularly detectable in the stories linked with the main protagonists: Munarra, a creature from the Dreaming, and Uncle, a Yorta Yorta elder. Munarra is first seen in Scene Two, "A River Is Made," in which she falls from the sky and "*cries a river*" (4): "Sounds of sobbing and mourning wreak out of the landscape. [...] the river is made" (5). Her presence on the stage bespeaks a magic realist conflation of ordinary reality and the supernatural. Two dingoes also drop from the sky in Scene Four. They explain how Munarra was sent to the earth by Biami, the creator, Munarra's husband, in order to lift the curse from the land. Like Ernestine in Tomson Highway's play, Munarra can be compared to a Saviour, a Christ-like figure. Thus, as in Highway's work, the land is strongly associated with female agency. These two dingoes—roles characteristically performed by actors in a blurring of boundaries between the human and the non-human—are supposed to protect Munarra. Ironically, they are afraid of every single noise they hear, which lends a farcical mood to the play. In Scene Eight, Munarra feels deeply wounded as she hears the crying of the Murray river—implicitly likened to a woman: "MUNARRA: No! Dhungula! Please stop crying! Please stop! No more crying! Stop it! I can't stand it anymore! You're breaking my heart" (18). The sense of fusion between the mythical Munarra and the river is manifest. This sense of loss recurs in Scene Thirteen, in which Munarra and the dingoes are "*surrounded by dead and dying trees. Inside each tree there is a spirit. The trees are intently watching MUNARRA and beckon to her to save them*" (26). Munarra bemoans the disappearance of the landmarks of the place. Towards the end of the scene, she hears the spirits of the deceased Indigenous people talking to her from under the earth in the Yorta Yorta language. She exclaims: "Sadness fills this place like a lake. I'm swimming in it" (28). In Act II, Scene Twenty, Munarra confronts the ghost of the first white settler on Yorta Yorta land, Sir Edward Curr. In a gesture of revenge against the confiscation of Yorta Yorta territories, she forces herself upon his hospitality and enjoys tea at his house. This moment of magic realism is thus endowed with marked political overtones. As she leaves Sir Edward Curr, Munarra symbolically places the teacup in her bag, a metaphor for the Aboriginal repossessing of the land: "CURR: Excuse me, madam. That teacup is mine. /MUNARRA: What's yours is mine" (42). Munarra's defiant stance culminates in her final assertion of Aboriginal presence, as she boldly says to Curr: "You will see us" (43). In Scene Twenty-five, Munarra prepares for war. Like her, the land is ready to fight: "*A storm is threatening as the land rises up in opposition. In full war regalia, MUNARRA and the DINGOES pick up their weapons.* [...] *MUNARRA screams to rid the land of this curse.* [...] *Weapons raised, we see the whites of MUNARRA's eyes. CURR petrified. The land is restless*" (52-53). In the ultimate scene, after Judge Olney's judgment is upheld and the land title is denied to Yorta Yorta people, Munarra nonetheless voices a last message of defiance. First, when seeing men digging her grave, she bursts out: "I see! You are driving us to distraction. Leading us to despair. You have taken. And now you give us death! [...] Are you waiting for my death? [...] You're digging my grave!" (53). At this point, she summons the power of the land and the Dreaming to avoid annihilation. Her outcry eventually forces the gravediggers to disappear:

> As long as this river flows, the lake fills and floods, the redgums sway and grow—we are here. For as long as the eagle flies and the long-neck turtle swims—we are here. Like every grain of sand on our river's shores—we are here. Like the line that connects me to my traditions—we are here. Like every leaf on every branch on every tree on this land—we are here. [...]
> I am not your servant.
> I am Yorta Yorta.
> And...
> WE...ARE...HERE. (53-54).

Munarra's spirit of endurance provides an apt coda to the entire play, subsuming both the political and ecological themes developed in the previous scenes.

Intertwined with Munarra's storytelling is the tale of Uncle, who is, we are told in the cast of characters, a Yorta Yorta elder who has returned to the river country to die. In the depiction of his last days on earth, Andrea James combines magic realism and storytelling. In Scene Nine, Uncle engages in a conversation with an old cod. This fishing episode emphasizes the profound ecological link between human and non-human realities, particularly as the cod bears a human name: "I'm gonna catch you today, Harold. You big, ugly Murray Cod, can you hear me? It's ya old mate Galnya here" (18). Uncle associates the cod with something superior to human life, something almost as old as a Dreaming creature: "You old dinosaur fish! You heard the songs travelling down the river" (19). The cod is not only linked to the history of the Murray river, it also knows the secret location of the "canoe tree," a source of spiritual inspiration to Yorta Yorta people: "That's sacred that tree. Don't you go telling anyone where it is, Harold!" (19). Uncle nonetheless shares this secret with Lyall, a young activist for land justice. In Scene Sixteen, as the two men approach the tree, "*Yorta Yorta spirits are seen and heard*" (33). In this magic realist atmosphere, the two men discover, much to their horror, that the canoe tree has been destroyed by white people. This evidences the desecration of Yorta Yorta culture by colonists: "LYALL *steps aside to reveal a tree stump and fresh sawdust on the earth.* UNCLE *and* LYALL *stare at it for a long time*" (33). This shocking discovery prompts Uncle to testify in the court hearings about land rights. However, the judge rejects his plea, as technically, Uncle was away from home for twenty years while placed in child welfare. The hearing reawakens the childhood wound of the loss of Uncle's little sister Mae, who drowned in the River while playing with her brother. However, Uncle asserts that in his heart, he never left the Aboriginal land of his ancestors (46). In Scene Twenty-three, Uncle finally catches old Harold. His speech to the cod reveals the ecological connotations associated with both fish and trees: "I knew I'd catch ya. Sooner or later. Later, rather than sooner. [...] You look to be three hundred year old [*sic*]. My dad said if you split open the lungs of a fish you can see its entire life journey mapped out. All the creeks and rivers the fish has swum. You see it all like a tree. That's your birth tree" (48). All life forms are undifferentiated in Nature. Now feeling he is about to die, Uncle relives a childhood experience when he and his sister Mae were lost in the bush. Uncle's past is unveiled to us as was the cod's through its tree-like lungs. In this episode, the spirits of the trees saved the children by providing the necessary light for them to walk home (50). In a present-day echo of this childhood scene, "*the wind changes its tune and MAE's voice leads him slowly into the darkness of the bush*" (48). Mae's spirit thus reunites Uncle with his Dreaming. Death is associated with the joy of coming home he had felt when, as a child, he was saved from the bush by the magical help of the trees. Uncle's demise suggests serenity, a kind of fusion with the land of the ancestors that contemporary Yorta Yorta people can only yearn for.

The aesthetics of storytelling and magic realism, rooted in Indigenous culture, are clearly associated with Munarra and Uncle. In marked contrast, Andrea James's historiographic agenda relies on the use of primarily Western theatrical and literary conventions. Sir Edward Curtis's exploration of Yorta Yorta land is recounted in Scene Five, ironically titled "The Curse." Curr quotes from his memoirs, which creates a kind of collage of literary and theatrical devices: "The party was rather a defenceless one, consisting principally of women, old men, and children, the young men being out hunting, so that on seeing us, a general stampede took place for the shore" (10). This sense of chaos is re-enacted by the dingoes and an Indigenous old man, who profusely insults Sir Edward Curr in the Yorta Yorta language (the printed script provides English translations). The Aborigines voice a threatening "Yanagai! Yanagai!" This bilingual aspect reintroduces Indigenous culture into a largely Western scenic presentation.

Further, the play is interspersed throughout with transcript-like scenes that document the proceedings of the Aboriginal land claims court hearings. These documentary-like episodes, conceived in a Western style, are nonetheless marked by the occasional use of Yorta Yorta language. For instance, when he testifies, Lesley insults the Judge to comic effect. Lesley exclaims: "Manarraupna mutja," the English translation of which, "Thunder in your anus!" is projected onto a backdrop screen. Lesley lies when he tells him that the words mean "You are my friend" (25). The judge remains unaware throughout that he is being ridiculed and that his presence on Yorta Yorta traditional land is being questioned. The court scenes potently articulate the incompatibility between Aboriginal oral culture and the Western reliance on written texts, particularly as Lesley says: "The truth is in the trees. [...] the truth is in the water, in the bush!" This character accordingly refuses to swear he will tell the truth (24). The conclusion of the play is likewise characterized by a kind of collage meant to express Yorta Yorta historical resistance against white people: "*A montage of images fill the space, emerging out of the land and onto the* TELLERS. *Yorta Yorta people past and present* [...] *The sound of a group of elders singing the song 'Bura Fera' echoes out in celebration of the endurance of Yorta survival and spirit*" (54). The use of Indigenous singing thus merges with the Western theatrical device of projected photographs. The final image of the play ironically re-articulates the spirit of Yorta Yorta culture, implying that this Indigenous community actually has the power to promulgate laws:

> This is an important
> ABORIGINAL SITE
> It is an offence to
> enter, deface, damage or
> otherwise interfere with
> this site
> Penalty:
> 1. Up to $ 10,000 or 5 years in jail or both
> 2. If the person is a body corporate, $ 50,000 (54)

This final statement recalls the defiance contained in the recitation of the Laurier Memorial in Highway's *Ernestine Shuswap Gets Her Trout*.

All in all, Andrea James's *Yanagai! Yanagai!* displays a hybrid combination of storytelling, magic realism and documentary transcripts recalling the tricksterish fluidity of Tomson Highway's aesthetics. Meant as the expression of a local Aboriginal Dreaming rooted in the Indigenous environment, Andrea James's stage idiom successfully articulates the power of Indigenous embeddedness in a globalized age.

III

Hone Kouka's *Waiora* (1996) is undoubtedly a classic of Māori playwriting. In this work, Kouka, whose lineage can be traced to Ngati Porou, Ngati Kahungunu and Ngati Raukawa tribes (*Playmarket Website*), creates a fluid, non-Aristotelian aesthetics relying on magic realism and Māori epistemologies. In this way, he subtly articulates the ecological link between Māori and their homeland. Māori people, being of Polynesian descent, are characteristically called the tangata whenua—the people of the land (Royal "Māori" 1). Māori cosmology presupposes a deep connection between humans and nature. It even posits a sacred relationship with the natural environment (Royal "Māori" 2). In *Waiora*, Kouka, like Highway and James, adopts a historiographic stance to express the uprootedness of a Māori whanau [family], who, in 1965, left their rural homeland (Te Ūkaipō) in the North Island to settle in an urban environment off the coast of the South Island. This severing of the link with the land of the ancestors provokes a crisis of displacement. Even though the father, Hone, makes good money by working in a sawmill, his family falls apart because it is deprived of its nurturing cultural roots. The youngest daughter, Rongo, misses Waiora so much that she almost dies. After realizing that he will never be totally accepted in Pākehā [Anglo-Celtic] society, Hone decides to retain the spiritual values of harmony associated with the homeland, Waiora. In this way, the conclusion suggests, the urban environment of the family can be endowed with a sense of place. Throughout the play, the presence of a chorus of Tīpuna [ancestors] suggests the contiguity between ordinary reality and the supernatural, as was the case in *Ernestine Shuswap Get Her Trout* and *Yanagai! Yanagai!* In her foreword, Māori writer Roma Potiki even suggests a link between the universe of the play and Māori mythology, arguing that the shadows of such gods as Tangaroa, Papatūānuku, Ranginui, and Tāne Mahuta haunt the displaced family (Potiki 10). In her afterword, Judith Dale confirms this when describing the set of the original production: "[…] a beautiful and luminously mirroring set of sea and sky was used to evoke the beach […] the designer went on to say that 'the parting of mirrors has loose references to Rangi and Papatuanuku and the creation of the world'" (Dale 117). Characteristically, Māori myths are rooted in the environment, as deities themselves are connected to natural elements. The Māori creation myth, which varies in its details according to local tribes, relies mostly on the separation between the sky and the earth. Originally, Ranginui, the Sky Father, and Papatūānuku, the Earth Mother, lay together. Eight of their children are particularly well-known: "Tangaroa (god of the sea), Tāne (god of the forests), Tāwhirimātea (god of winds and elements), Rongomatāne (god of the kūmara and cultivated crops), Haumia (god of the fernroot and wild fruits and herbs), Tūmatauenga (god of man and war), and Whiro (god of evil, disease, and pestilence)" (Barlow 10). These child gods felt cramped in the close embrace between Rangi and Papa. Tāne finally succeeded in separating their parents, by throwing Rangi upwards. Thus, the children were given light (Sinclair 11; Barlow 11). Of particular importance in the ecological imaginary of *Wairoa* is the liminal space of the beach, the threshold between the world of the living and that of the ancestors. This special kind of space is closely related to Māori myths. Indeed, the spirits of those who had died plunged into the sea at Cape Reinga in the far North and relocated to "a place where the spirits of humans resided before coming to this life [...] known as Hawaiki" (Barlow 111). As suggested in this myth, the Māori notion of time differs from the linear vision inherent in Western culture. Scholars often compare Māori time to a spiral, which evokes circularity and variation. The spiral is pervasive in Māori wood carvings, particularly in the form of a koru [fern], an artistic rendition of the link between Māori culture and Nature. As Charles Royal notes, "The koru, which is often used in Māori art as a symbol of creation, is based on the shape of an unfurling fern frond. Its circular shape conveys the idea of perpetual movement, and its inward coil suggests a return to the point of origin. The koru therefore symbolizes the way in which life both changes and stays the same" (Royal "Māori creation traditions"). The motif of the koru, suggesting both change and continuity, is particularly apt to depict the final healing characterizing *Waiora*. This ending recreates a kind of family unity similar to that experienced in the place of origins, albeit in a less congenial setting. Further, as it clearly departs from Aristotelian linearity, Kouka's magic realist aesthetics could metaphorically be compared to the spiral shape of the koru. These patterns bring to mind the traditional Māori carvings found on the poles of a whare whakairo [carved ancestral house]. In this sense, Kouka's "ecological" aesthetics, like Highway's and James's, offers a markedly local idiom opposing the tendency towards sameness inherent in globalization.

The play's prologue boldly articulates this non-linear aesthetics. While a number of scenes are written in a deceptively Western naturalistic manner, this introduction foregrounds a magic realist style that shows affinities with what Jeanne Delbaere calls "mythic realism," a form of magic realism rooted in the environment and, I would further contend, in Indigenous mythologies (Delbaere 252-53). To emphasize the "magic" aspect inherent in Delbaere's concept, I shall use the phrase "mythic magic realism" throughout this section. Indeed, in this prologue, the family's Tīpuna, a chorus of four actors, assert their strong influence on the family as soon as they appear on the beach. This liminal space, evocative of the Māori God of the sea, Tangaroa, thus clearly conjoins the world of the living and of the spirits.[8] The latter's presence is portentous, as indicated in the stage directions: "*After surveying the place, they quietly begin to sing the waiata 'Taukuri E'. It is slow and mournful, talking of the pain of leaving the papa kāinga* [homeland]" (15). Their waiata [song] and their subsequent haka [dance] introduce typically Māori forms of ritual and art into the play. In addition, the waiata explicitly alludes to the Māori creation myth, invoking Ranginui and Papa in its elegiac yearning for a lost sense of place:

> Taukuri e… Aue rā, e te hau kāinga
> Mākūkū ana i aku roimata
> Taukuri e… Te ūnga o te tāngaengae
> E kukume noa nei te ate
> Ranginui… E tū nei titiro kau iho
> Ki ahau e whakarere atu nei
> I a Papa… Kia kāewaewa noa rā
> Kia tika ai ko te kimi oranga! (15)
>
> [Alas… I cry for my place of origin
> Now left saturated in my tears
> Alas… the place where my umbilical cord is still
> Attached and pulls at my heart with almighty
> Strength
> Ranginui… Bear witness upon me as I leave behind all I
> Know, all I am, including…
> Papa… Land walked upon by generations many before
> Me As I now wander (seemingly) aimlessly in
> Search of well-being. (110)]

This lament, although addressed to Rangunui, the Sky Father, and Papa, the Earth Mother, seems to be directed especially towards Rongo, the youngest daughter of the family, who is turning eighteen as the play opens. While she epitomizes the family's crisis of identity, she also brings about the final sense of regeneration. As Judith Dale indicates in her afterword, Rongo's name means several things: "peace," "to listen, to hear" and "news" (Dale 115). This symbolically refers to Rongo's role as someone who "listens" to ancestral spirits. In doing so, she enables renewal. As the prologue draws to a close, the entire whānau [family] and the Tīpuna join in a highly stylized haka, a symbol of the fusion between mundane and spiritual realities (16). Sporadic scenes of comparable mythic magic realism alternate with naturalistic episodes at regular intervals during the play, in a spiral-like movement evocative of a koru design.

---

8 In her introduction to the play, Roma Potiki affirms that the presence of Tangaora can be felt throughout the play (Potiki 10).

Act I dramatizes the celebration of Rongo's coming of age. Hone's boss, the Pākehā Steve Campbell, is invited to the party. Hone, who feels privileged to work at the sawmill, secretly hopes his manager will promote him to the rank of foreman. In stark contrast to this festive mood, Rongo sings a plaintive waiata at the beach, where she is observed by her Tīpuna. She bemoans the loss of connection with the land of origin, again evoking Māori mythology: "Far away is Hawaiiki [...]. How lonely my heart aches for those ways of old/and the source of revitalising energy/the being I once was, now buried inside me" (29; translated 111). In a subsequent monologue in English, addressed to her dead grandmother, she reinforces this sense of frustration. She refers to Tangaroa, the god of the tides, establishing a connection between Māori deities, Nature, and home:

> I am standing in the water so I can touch home. Kei te whānui ngā ringaringa o Tangaroa hei awhi a Papatūānuku [The hands of Tangaroa reach out along Papatūānuku]. His hands smooth out the sand you see. [...] So if I am held in those hands, I am taken back to the beach of Waiora, our true home., The Homeland. Waiora Te Ūkaipō. [...] Here, we kōrero Pākehā [we speak English] [...]. We are stopping ourselves from speaking the reo [Māori language]. [...] We will be a lost people [...]. Lost [...] Ka ngaro [Lost]. (30-31)

The chorus of ancestors then urges her to follow them in the spirit world. Rongo feels that her death would bring together members of her family, which is threatening to fall apart. Indeed, as we learn in the following scenes, Hone's first son, Mahurangi, left the family after arguing with his father (46). In Act II, we are told that the youngest son, Boyboy, cannot assimilate at the local Pākehā school. He is ultimately dismissed on false allegations of striking a racist teacher (100). Shortly after Rongo's lament, the Tīpuna enjoy upsetting the family by starting a fire, which creates havoc: "E...i tēnā [Burn...fire...burn]" (42; transl. 112). In a trickster-like spirit recalling the Māori legendary hero Maui, "*The* Tīpuna *are laughing at what they have caused*" (43). Durix reminds us of the tricksterish abilities of this Māori hero, who was known for "fishing up new land from the bottom of the sea" (Durix 436). Through this veiled allusion to a trickster-like demi-god, *Waiora* echoes Highway's *Ernestine Shuswap Gets Her Trout*.

Steve Campbell, the Pākehā manager, at first seems to enjoy a relationship to the land similar to that of the Māori characters: "This place owns me and my family. My grandfather's buried here, so we're always part of it now. The land. [...] It's the land, ya understand that ...the bond with the land" (64-65). In Act II, however, Steve's materialistic vision is evidenced when he dismisses the Māori ways of cultivating the land: "Yeah, if you run the country, to be honest, it'd be a bit of shambles. Since we got here, we've been building this place for a better future" (84). Steve later confesses that he could not allow Hone to become a foreman, precisely because he is Māori (89). He simply intends to give him a monetary bonus, a gesture epitomizing his lack of spirituality. When realizing his illusions, Hone bemoans the persistent racism inherent in Pākehā society: "Things haven't changed that much" (98). At the same time, the Tīpuna again try to win Rongo over. They sing a mournful waiata: "Auē, taukuri e…kua wheturangihia koe…/E Rongo e…okioki, e moe. [Alas, my child…you have become/one of the night stars/Oh Rongo…rest now, sleep]" (104; transl. 113). Again, the Tīpuna are acting in a tricksterish way, as "*they are toying with the family*" (105).

The epiphanic conclusion of the play offers a successful reenactment of the Māori creation myth, which yields ecological connotations: it enables the family to recapture the harmony associated with Waiora. In this final episode, the similarities between Hone and the god Tāne are at their most striking. As a sawmill employee, Hone seems endowed with Tāne's sway over the forests, although he ironically destroys them by adhering to Pākehā commercial ethics. However, Hone also resembles Tāne in a much more positive manner, as towards the end of the play, he behaves according to Māori spiritual values. Indeed, Tāne managed to separate Rangi, the Sky Father, from Papa, the Earth, thus leading the other children to a world of light. In an echo of this mythical event, Hone decides to confront the Tīpuna, in order to prevent them from killing Rongo: he "*begins a haka of self-hate, self-loathing and remorse*" (107). In a gesture of reconciliation with his father, Boyboy joins his father's haka: "*The two fight for* RONGO" (109). They eventually succeed in preventing the Tīpuna from leading Rongo to Hawaiiki. Serenity is achieved as Rongo "*has heard her family. She has come back. The family have beaten the Tīpuna*" (109). This climax, which is written in the same mythic magic realist vein as the play's prologue, foregrounds a metaphorical reinscription of the Māori homeland into the family's new urban environment. If memories of belonging are preserved, this process nonetheless implies a variation: the family will now be more cohesive than ever, in an effort to withstand the woes of migration. This blend of continuity and change thus brings to mind the spiral-like motif of the koru. In addition, as mentioned before, the non-linearity of Kouka's use of mythic magic realism, alternating with naturalistic passages, is deeply connected to the carving motif of the koru. All in all, in *Waiora*, Kouka develops an innovative shape-shifting aesthetics rooted in Māori culture. This style enables him to couch his ecological vision of the world in a vigorously local idiom, one that evades homogenizing globalization.

***

Summing up, *Ernestine Shuswap Gets Her Trout*, *Yanagai! Yanagai!*, and *Waiora* share a number of transcultural characteristics. In these plays, ecological thinking is primarily associated with water. The

Pacific Ocean and the Thompson and Murray rivers literally create the spiritual sense of place embedded in Indigenous land. Moreover, these plays rely on a historiographic mode of presentation, thus suggesting the profound link between the (post)colonial past and ecological issues. Clearly, the pivotal role of history in these works shows how closely related postcolonial studies and ecocritcism are. However, Highway, James and Kouka articulate these thematic similarities in a unique aesthetic idiom. Their plays provide three contrasting ways in which Indigenous dramatists from various parts of the world have negotiated the numerous possiblities of expressing a web of connections between the human and the non-human. Their works highlight the shape-shifting diversity of Indigenous ecological aesthetics, while offering insights into different cosmologies. Much further comparative scholarly work will be required, if ecocritics wish to fully document the role these innovative aesthetics perform in preserving localized knowledges in today's gobalized world.

Works Cited

Aarons, Wendy and Theresa J. May, eds. *Readings in Performance and Ecology*. New York: Palgrave Macmillan, 2012.

Barlow, Cleve. *Tikanga Whakaaro. Key Concepts in Māori Culture*. Oxford: Oxford UP, 1991.

Buell, Lawrence. *The Future of Environmental Criticism. Environmental Crisis and Literary Imagination*. Oxford: Blackwell Publishing, 2005.

Caruana, Wally. *Aboriginal Art*. London and New York: Thames and Hudson, 1993.

Clark, Timothy. *The Cambridge Introduction to Literature and the Environment*. Cambridge: Cambridge UP, 2011.

Dale, Judith. "Afterword." *Waiora*. Hone Kouka. 115-21.

Delbaere, Jeanne. "Psychic Realism, Mythic Realism, Grotesque Realism: Variations on Magic Realism in Contemporary Literature in English." Ed. Lois Parkinson Zamora and Wendy B. Faris. *Magical Realism. Theory, History, Community*. Durham and London: Duke UP, 1995. 249-63.

De Wagter, Caroline. *"Mouths on Fire with Songs." Negotiating Multi-ethnic Identities on the Contemporary North American Stage*. Amsterdam and New York: Rodopi, 2013.

Durix, Jean-Pierre. "Patricia Grace's *Potiki* or the Trickster Behind the Scenes." *A Talent(ed) Digger. Creations, Cameos, and Essays in Honour of Anna Rutherford*. Ed. Hena-Maes Jelinek, Gordon Colllier, and Geoffrey V. Davis. Amsterdam and New York, Rodopi, 1996. 436-41.

Garrard, Greg. *Ecocriticism*. London and New York: Routledge, 2012.

George, Miria. *Urban Hymns*. In *Three Young & Hungry Plays*. Wellington: Playmarket, 2010. 119-77.

Gilbert, Helen, ed. *Postcolonial Plays. An Anthology*. London: Routledge, 2001.

Gilbert, Helen and Joanne Tompkins. *Post-colonial Drama. Theory, Practice, Politics*. London and New York: Routledge, 1996.

Gray, Nelson. "Bring Blood to Ghosts: English Canadian Drama and the Ecopolitics of Place." Ed. Wendy Arons and Theresa J. May. *Readings in Performance and Ecology*. 23-32.

Highway, Tomson. *Ernestine Shuswap Gets Her Trout*. Vancouver: Talonbooks, 2005.

—. *The Rez Sisters*. Saskatoon: Fifth House, 1988.

Huggan, Graham and Helen Tiffin. *Postcolonial Ecocriticism*. London and New York: Routledge, 2010.

James, Andrea. *Yanagai! Yanagai!* Sydney: Currency Press, 2003.

Kouka, Hone. *Waiora*. Wellington: Huia Publishers, 1997.

Maufort, Marc. *Transgressive Itineraries. Postcolonial Hybridizations of Dramatic Realism*. Brussels: P.I.E.-Peter Lang, 2003.

Playmarket Website. Web. 12 November 2013. http://www.playmarket.org.nz/playwrights/hone-kouka.

Potiki, Roma. "Foreword." *Waiora*. Hone Kouka. 8-11.

Royal, Te Ahukaramū Charles. "Māori." *Te Ara-The Encyclopedia of New Zealand*, updated 21 August 2013. Web. 30 October 2013. http://www.TeAra.govt.nz/en/maori.

—. "Māori Creation Traditions-Common Threads in Creation Stories." *Te Ara-The Encyclopedia of New Zealand*, updated 15 November 2012. Web. 30 October 2013. http://www.TeAra.govt.nz/en/photograph/2422/the-koru.

Sabbioni, Jennifer. "Preface." *Indigenous Australian Voices. A Reader*. Ed. Jennifer Sabbioni, Kay Schaffer, and Sidonie Smith. New Brunswick, NJ, and London: Rutgers UP, 1998. xix-xxxi.

Sayers, Andrew. *Australian Art*. Oxford: Oxford UP, 2001.

Sinclair, Keith. *A History of New Zealand*. Auckland: Penguin Books, rev. ed. 2000.

Stanlake, Christy. *Native American Drama: A Critical Perspective*. Cambridge: Cambridge UP, 2009.

Ward, Russel. *Concise History of Australia*. St Lucia: U of Queensland P, 1965.

Notes on Contributors

Maryrose Casey is an Australian Research Council (ARC) Future Fellow with Monash Indigenous Centre, Monash University. Her major publications include *Creating Frames; Contemporary Indigenous Theatre* (2004), *Transnational Whiteness Matters* (2008), co-edited with Aileen Moreton-Robinson and Fiona Nicoll, and *Telling Stories: Aboriginal and Torres Strait Islander Theatre Practices* (2012).

Jaye T. Darby is a transformative scholar, activist, and educator. With Hanay Geiogamah, she is co-founder and co-director of Project HOOP, an initiative to advance Native performing arts artistically, academically, and professionally. She is the co-editor of the anthology *Keepers of the Morning Star: An Anthology of Native Women's Theater* with Stephanie Fitzgerald. With Geiogamah, she recently co-edited *American Indian Performing Arts: Critical Directions*, which draws on Indigenous perspectives. She also teaches in the Graduate School of Education and Information Studies at the University of California, Los Angeles.

Birgit Däwes is Professor and Chair of American Studies at the University of Vienna. Having taught previously at the Universities of Mainz and Würzburg as well as at National Sun Yat-sen University in Kaohsiung, Taiwan, she is a specialist in Native North American literatures and cultures. Next to a monograph study on *Native North American Theater in a Global Age* (2007) and the recent collection, *Indigenous North American Drama: A Multivocal History* (2013), she has also published an award-winning study on fiction as a mode of cultural memory: *Ground Zero Fiction: History, Memory, and Representation in the American 9/11 Novel* (2011).

Nicholle Dragone, an Indigenous scholar of Lakota/Dakota descent, teaches American Indian Studies and English at Black Hills State University. Dragone completed her doctoral degree in American Studies and Law Degree at the University at Buffalo. She has published articles and book chapters on Native American literature and is currently editing a critical anthology on Onondaga writer Eric Gansworth's literary and visual art.

Hilary Halba is Senior Lecturer and Head of Programme in Theatre Studies at the University of Otago in Aotearoa/New Zealand. She has had a long association with Māori and Bicultural theatre in Te Wai Pounamu (the South Island of New Zealand), and has directed and researched this form of theatre in partnership with Māori for over 15 years. Hilary is also an award-winning actor, having studied acting and the teaching of acting at the prestigious Neighborhood Playhouse in New York City. More recently, she has embarked on creating practice-led verbatim theatre projects with her colleague Stuart Young.

Maryann Henck has been teaching literature and culture seminars at Leuphana Universität Lüneburg in Germany since October 2000. She holds a Master of Arts in Germanic Languages and Literature from the University of Pennsylvania. Together with her colleagues, Maria Moss and Sabrina Völz, she co-edited the book, *White-Indian Relations: Moving into the 21st Century*. In the winter terms 2009/2010 and 2011/2012, she organized Canadian/Anishnawbe author Drew Hayden Taylor's stays as writer-in-residence as well as staged adapted versions of his plays, *In a World Created by a Drunken God* and *Dead White Writer on the Floor*.

Ric Knowles is Professor of Theatre Studies at the University of Guelph, Editor of *Theatre Journal*, and from June 2013 to July 2014 a Fellow at the International Research Centre "Interweaving Performance Cultures" at the Freie Universität, Berlin. He is the award-winning author of six and editor of eleven books on theatre and performance. His latest authored books are *Theatre & Interculturalism* (2010) and *How Theatre Means* (forthcoming, 2014). He is also an active professional dramaturge who focuses on inter- and cross-cultural performance.

Diana Looser received her Ph.D. in Theatre Arts from Cornell University, and is currently a Lecturer in Drama at the University of Queensland, Australia. Her research interests include postcolonial, transnational, and intercultural theatre and performance, with a particular focus on the Pacific Islands region. Her essays have appeared in several journals and edited collections, and she is the author of the monograph *Remaking Pacific Pasts: History, Memory, and Identity in Contemporary Theatre from Oceania*, to be published by the University of Hawai'i Press.

Marc Maufort is Professor of English, American and postcolonial literatures at the Université Libre de Bruxelles (ULB, Belgium). He is currently one of the Vice-Presidents of the International Association for Comparative Literature (ICLA). Maufort has written and (co)-edited several books on O'Neill as well as postcolonial and multi-ethnic drama, including *Transgressive Itineraries: Postcolonial Hybridizations of Dramatic Realism* (2003), and *Labyrinth of Hybridities: Avatars of O'Neillian Realism in Multi-ethnic American Drama* (1972-2003) (2010).

Yvette Nolan is a playwright, dramaturg, and director. Her plays include *Annie Mae's Movement*, *BLADE*, *Job's Wife*, *Ham and the Ram*, *Scattering Jake*, *The Birds* (a modern adaptation of Aristophanes' comedy) and *The*

Unplugging. She is the editor of *Beyond the Pale: Dramatic Writing from First Nations Writers and Writers of Colour*, and the co-editor, with Donna-Michelle St Bernard, of *Refractions: Solo*. From 2003-2011, she served as the Artistic Director of Native Earth Performing Arts in Toronto. She is currently working on *Medicine Shows*, a book about Native theatre in Canada.

David O'Donnell is a theatre director and Associate Professor in Theatre at Victoria University, Wellington, Aotearoa/ New Zealand. David has directed many premieres of New Zealand plays, most recently *Te Karakia* by Albert Belz, *Collapsing Creation* by Arthur Meek and *West End Girls* by Ken Duncum. He has published widely on New Zealand and Australian theatre, Māori and Pacific performance and community theatre. With Marc Maufort, he co-edited *Performing Aotearoa* (2007) and is currently editor of the Playmarket New Zealand Play Series.

Ginny Ratsoy, Associate Professor of English and Provost's Fellow of Teaching and Learning Support at Thompson Rivers University, edited *Theatre in British Columbia,* and co-edited, with James Hoffman, *Playing the Pacific Province: An Anthology of British Columbia Plays, 1967-2000.* She has also written chapters and articles on culture in small cities, Canadian drama, and transformative higher education.

Rachael Swain is a founding member and Director of Stalker Theatre based in Sydney. She conceives and directs collaborative multimedia works creating a unique synthesis of dance, theatre, circus and video. Rachael is also a founder and Co-Artistic Director (with Dalisa Pigram) of Marrugeku, the innovative intercultural-Indigenous dance theatre company based in Broome in the far north of Western Australia. She conceived and directed *Mimi* (1996), *Blood Vessel* (1998), *Crying Baby* (2000), *Incognita* (2002) co-directed with Koen Augustijnen, *Burning Daylight* (2006), *SUGAR* (2007), *Shanghai Lady Killer* (2010) *and Buru* (2010) co-directed with Dalisa Pigram.

Lisa Warrington is an Associate Professor in Theatre Studies at the University of Otago, New Zealand. She is also a theatre director of many years standing, with over 130 productions to her credit. She is a long-time champion of New Zealand theatre. Her current research interests include Pasifika theatre, Shakespeare in performance, 19th century New Zealand theatre and directing. She created and runs Theatre Aotearoa, a major database which covers theatre productions in New Zealand from the early 19th century to the present, which can be found at http://tadb.otago.ac.nz.

Index

Dramaturgies

Texts, Cultures and Performances

This series presents innovative research work in the dramaturgies of the twentieth and twenty-first centuries. Its main purpose is to re-assess the complex relationship between textual studies, cultural and/or performance aspects at the dawn of this new multicultural millennium. The series offers discussions of the link between drama and multiculturalism (studies of "minority" playwrights—ethnic, Aboriginal, gay, and lesbian), reconsiderations of established playwrights in the light of contemporary critical theories, studies of the interface between theatre practice and textual analysis, studies of marginalized theatrical practices (circus, vaudeville, etc.), explorations of emerging postcolonial drama, research into new modes of dramatic expressions and comparative or theoretical drama studies.

*The Series Editor, **Marc Maufort**, is Professor of English literature and drama at the Université Libre de Bruxelles.*

Series Titles

N° 33 – Birgit Däwes and Marc Maufort (eds.), *Enacting Nature. Ecocritical Perspectives on Indigenous Performance*, 2014, ISBN 978-2-87574-146-2.

N° 32 – Gerald Gillespie, *Ludwig Tieck's* Puss-in-Boots *and Theater of the Absurd. A Commentated Bilingual Edition*, 2013, ISBN 978-2-87574-026-7.

N° 31 – Julie Vatain, *Traduire la lettre vive. Duos anglais sur la scène fran*çaise, 2012, ISBN 978-2-87574-012-0.

N° 30 – Ève Feuillebois-Pierunek (dir.), *Théâtres d'Asie et d'Orient. Traditions, rencontres, métissages*, 2012, ISBN 978-90-5201-847-8.

N° 29 – Thierry Dubost (ed.), *Drama Reinvented. Theatre Adaptation in Ireland (1970-2007)*, 2012, ISBN 978-90-5201-800-3.

N° 28 – Dorothy Figueira and Marc Maufort (eds.), with the assistance of Sylvie Vranckx, *Theatres in the Round. Multi-ethnic, Indigenous, and Intertextual Dialogues in Drama*, 2011, ISBN 978-90-5201-690-0.

N° 27 – Sébastien Ruffo, *Jeux d'acteurs comparés. Les voix de Belmondo, Depardieu, Lebeau et Nadon en Cyrano de Bergerac*, 2011, ISBN 978-90-5201-657-3.

N° 26 – Catherine Bouko, *Théâtre et réception. Le spectateur postdramatique*, 2010, ISBN 978-90-5201-653-5.

N° 25 – Marc Maufort, *Labyrinth of Hybridities. Avatars of O'Neillian Realism in Multi-ethnic American Drama (1972-2003)*, 2010, ISBN 978-90-5201-033-5.

N° 24 – Marc Maufort & Caroline De Wagter (eds.), *Signatures of the Past. Cultural Memory in Contemporary Anglophone North American Drama*, 2008, ISBN 978-90-5201-454-8.

N° 23 – Maya E. Roth & Sara Freeman (eds.), *International Dramaturgy. Translation & Transformations in the Theatre of Timberlake Wertenbaker*, 2008, ISBN 978-90-5201-396-1.

N° 22 – Marc Maufort & David O'Donnell (eds.), *Performing Aotearoa. New Zealand Theatre and Drama in an Age of Transition*, 2007, ISBN 978-90-5201-359-6.

N° 21 – Johan Callens, *Dis/Figuring Sam Shepard*, 2007, ISBN 978-90-5201-352-7.

N° 20 – Gay McAuley (ed.), *Unstable Ground. Performance and the Politics of Place*, 2006 (2e tirage 2008), ISBN 978-90-5201-036-6.

N° 19 – Geoffrey V. Davis & Anne Fuchs (eds.), *Staging New Britain. Aspects of Black and South Asian British Theatre Practice*, 2006, ISBN 978-90-5201-042-7.

N° 18 – André Helbo, *Signes du spectacle. Des arts vivants aux médias*, 2006, ISBN 978-90-5201-322-0.

N° 17 – Barbara Ozieblo & María Dolores Narbona-Carrión (eds.), *Codifying the National Self. Spectators, Actors and the American Dramatic Text*, 2006, ISBN 978-90-5201-028-1.

N° 16 – Rachel Fensham, *To Watch Theatre. Essays on Genre and Corporeality*, 2009, ISBN 978-90-5201-027-4.

N° 15 – Véronique Lemaire, with the help of/avec la collaboration de René Hainaux, *Theatre and Architecture – Stage Design – Costume. A Bibliographic Guide in Five languages (1970-2000) / Théâtre et Architecture – Scénographie – Costume. Guide bibliographique en cinq langues (1970-2000)*, 2006, ISBN 978-90-5201-281-0.

N° 14 – Valérie Bada, *Mnemopoetics. Memory and Slavery in African-American Drama*, 2008, ISBN 978-90-5201-276-6.

N° 13 – Johan CALLENS (ed.), *The Wooster Group and Its Traditions*, 2004, ISBN 978-90-5201-270-4.

N° 12 – Malgorzata BARTULA & Stefan SCHROER, *On Improvisation. Nine Conversations with Roberto Ciulli*, 2003, ISBN 978-90-5201-185-1.

Peter Lang - The website

Discover the general website
of the Peter Lang publishing group

www.peterlang.com

Imprimé en France
FROC031825281020
25563FR00006B/50

9 782875 741462